Acknowledgements

I express my sincere and deepest gratitude to my distinguished and revered Supervisor, Prof. DD Sharma, Chairman, Department of Geography, HPU, Shimla, under whose guidance and supervision the present study was undertaken and completed.

I would like to extend my thanks to Prof. BS Marh, Dr. PD Bhardwaj, Dr. Anurag Sharma and Dr. BR Thakur, faculty members department of Geography, for their kind help and valuable suggestions.

I also thanks the non-teaching staff of the department specially Dr. Sushil Kumar and Ratan for their cooperation during the research.

I am very much grateful to my parents / family members Sh. Junda Ram & Smt. Sayano Devi, Sh. M.L. Chauhan & Smt. Santosh Devi; Sh. Kumbiya Ram, Saniya Ram, Ashish Chauhan, Mamta, Pankaj and Jaipal who have always been a driving force behind all my achievements.

I am also thankful to my friends specially Dr. Jitender Kumar Sahni, Promila, Chhering, Rajender, Devender, Harish, Krishan, Chander Pal, Ashok, Nishant, Attar, Khyal, Dhan Dev, Ritesh, Amit, Kapil and Janam for their support and help during the research work.

Last but not the least I am greatly indebted to Mrs. Seema Chauhan, who always inspired and supported me at every time.

(Jagdish Chand)

TABLE OF CONTENTS

LIST OF TABLES

LIST OF FIGURES

LIST OF PHOTO PLATES

CHAPTER - I
INTRODUCTION

1.1 INTRODUCTION

Resources are the entities that are beneficial to human beings. They may be physical substances or human resources. To be considered a resource the entity should be amenable to use at the prevailing levels of technology. Even physical entities that may be in existence in a particular area but that cannot be utilized are not considered resources. Although physical entities are very important, they cannot be put to any use without the human or technological resources. Therefore, it is often emphasised that resources are not but they become. In other words, it is the human need and the capability to put them to use that make various entities the resources (Zimmerman, 1951).

The term 'natural resource' has undergone an expansion in meaning as a result of man's greater understanding of his relationship with the world he lives in. In the early twentieth century, the natural resources were viewed primarily as resources of valuable and useful commodities. They were mostly the raw materials present in the environment that man could use to some purpose, e.g. minerals and fuels, forests and grazing resources, wildlife and fishes. More recently the concept of a natural resource has been broadened to include the total natural environment, that is, the entire surface layer of the earth, because all parts of the earth's surface are of some use to man in that they contribute to the production of necessities and amenities that people demand. Thus from this point of view all living and non-living elements of the atmosphere, the oceans, the deserts, the polar, tropical and temperate regions, and the interior of the earth too have all become valuable resources. All these resources are to be utilized scientifically and managed with care to provide necessities and comforts to the present and future generations (Ramande, 1984).

Natural resources provide the base on which the edifice of development is raised. Its use depends upon the type of economy, the level of technology and preferences of the culture of a given society. The importance of natural resources is more critical to societies which are at a relatively low of development. People have to conform their livelihood and life style to the settings of nature. The sustainable use of natural resources to attain high levels of human development has become imperative. Natural resources of any places have a direct relationship with their physiographic conditions including relief, drainage, climate and geology. These in turn influence the type of soils and the kind of vegetation cover.

Forests, which constitute an important resource and activities based on to utilization of this resource provide employment to a large number of people. But in the modern era this activity has substentially declined. The term forestry refers to obtaining of various types of products from forests. It includes not only the production of timber but also the activities of gathering of tree products. Production of timber is the most advanced among the forestry related activities, and gathering of various products is one of the oldest human occupations. While gathering is a form of primitive subsistence activity, lumbering or production of timber is a modern method of utilization of forest resources. Gathering of forest products is generally the way of life of the people with a low level of cultural and economic development (Singhal et al. 2003).

Forests provide a multiplicity of environmental services. Foremost among these is the recharging of mountain aquifers, which sustain our rivers. They also conserve the soil, and prevent floods and drought. The forests provide habitat for wildlife and the ecological conditions for maintenance and natural evolution of genetic diversity of flora and fauna. They are the homes of traditional forest dependent communities and yield timber, fuel wood and other forest produce. The forests possess immense potential for economic benefits, in particular for local communities, from sustainable eco-tourism. On the other hand, in recent decades, there has been significant loss of forest cover, although there are no clear signs of reversal of this trend. The principal direct cause of forest loss has been the conversion of forests to agriculture, settlements, infrastructure, and industry. In addition, commercial extraction of fuelwood, illegal felling, and grazing by the cattle, has degraded forests. These causes, have their origins in the fact that the environmental values provided by forests are not realized as direct financial benefits by various parties, at least to the extent of exceeding the monetary incomes from alternative uses, including those arising from illegal use (Shafi and Raza, 1992).

Moreover, the antiquity forest dwelling communities had generally recognized traditional community entitlements over the forests. On account of this they had strong incentives to use the forests sustainably and to protect them from encroachers.

After the commencement of formal forest laws and institutions in 1865, these entitlements were effectively extinguished in many parts of the country. Such disempowerment has led to the forests becoming open access in nature, resulting to their gradual degradation in a classic manifestation of the tragedy of the commons. However, large scale forest loss would lead to catastrophic, permanent change in the country's ecology. It will definitely cause to major stress on water resources and soil erosion, with consequent loss of agricultural productivity, industrial potential, living conditions, and the onset of natural disasters like drought and floods. In any event, the environmental values of converted forests must be restored, as nearly as may be feasible, to the same publics. The National Forest Policy, 1988; and the Indian Forest Act, 1927; as well as the regulations under it, provide a comprehensive basis for forest conservation. The National Forest Commission, set up in 2003, is reviewing the policy, legislative and institutional basis of forest management. Nevertheless, it is necessary to further for the underlying causes of forest loss, to take necessary steps (Khullar, 2010).

India is a country where nearly 70 percent of its population lives in rural environment. She has 17.31 percent of the world's human population according to census 2011 and 15.1 percent of cattle population while it occupies 2.47 percent of the world geographical area and has only 1 percent of the world forests (Annual Report, 2012-13). Over the last four decades, there has been a large increase in not only human population but also in livestock population. The average exponential growth rate for 2001-2011 has declined to 1.64 percent per annum from 1.97 percent per annum during 1991-2001. The average annual exponential growth rate during 1981-1991 was 2.16 (Census of India, 2011). Similarly, the livestock population, which was 292 million in 1951, increased to 429.6 million in 1987 and must have crossed 500 million in 2000 AD (Dwivedi, 1993). India, with only 2.29 percent of the land area of the world, is maintaining about 10.71 percent of the world's livestock population (Annual Report, 2012-13). The steep increase in human and livestock population has put unprecedented demand on the natural resources of the country.

In addition, the forests have suffered serious depletion over the years, in terms of their growing stock, forest cover, creaming out of important timber. India's total forest area accounts for 685790 sq.km. in 2011. Out of which the share of Himachal Pradesh is 37033 sq. km which is 66.52 percent of its geographical area that is

4.83 percent, of total forest area of the country (SFR, 2011). In the recent years, this valuable resource is being over exploited without caring for the scientific principles that need to be taken into account while dealing with the extraction of such a resource. According to the forest survey of India (1996) report, the total requirement of firewood in the country is around 201 million tonnes. Out of this roughly about 103 million tonnes comes from the forest area (including plantations) which constitutes nearly 51 percent of the total requirements, while the balance 98 million tonnes comes from farm forestry sector including common lands (SFR, 2001). The FSI has estimated that the incremental growth of India's forests in terms of fuel wood is around 21million tonnes per year. Of this, around 17 million tonnes are available from forests at sustained basis. In this way, nearly 86 million tonnes of fuel wood is being removed from the forests and plantation of India every year in excess of what they are capable of producing on sustained basis. The study has also stated the share of fuel wood in total energy consumption increased from 54.57 percent to 61.10 percent for the periods of 1978-79 to 1992-93 (SFR, 2001). This however continues to be removed from the forest areas clearly indicating that they continue to be over exploited and consequently undergo degradation. This excesses exploitation gives rise to certain ecological problems. These ecological problems manifest themselves in the loss of green cover, soil erosion, flooding, loss of moisture, landslides, climatic changes etc. and as such have the negative impact on the very existence of human and other biota of the region.

Changes in forest ecosystem however, also take place due to geological and biological evolution and climatic changes. But changes, which occur in them as a result of human interaction, can be both destructive and regenerative. Fortunately, it is possible to regenerate forest resources. Thus by using natural and artificial method of forest regeneration, it is possible to maintain a flow of maximum return from forest resources on a sustained basis. Excessive grazing by domestic stock is another major reason for shrinking of forests and degradation of grazing and pasturelands. According to forest survey of India (1996) the demand of fodder in the year 2000 was estimated to be 2085 million tonnes. However, forests contributed 250 million tonnes of green and 441million tonnes of dry fodder during this year. Therefore, the total share of forest in the supply of fodder comes to be 691 million tonnes, which is about 1/3 percent of the total fodder requirements. The sustained production of fodder from forests is only about 45 million tonnes. Thus, forest areas are over grazed and the productivity keeps on declining (SFR, 2011). The forests, therefore, play a very significant role not only by meeting the physical requirements of the people at large but also by controlling physical environments too. Besides forest satisfy the aesthetic sense of the mankind. Moreover, the relationship of forest with air, water and soil is very complex. It is realized that over utilization of forest results in the lowering of the productivity of land, decrease in the surface water supply in the springs, streams, wells and diminution of underground water reservoirs associated with shrinking of water table. Further denudation of forests under these conditions leads to ecological and socio- economic desertification.

Since biodiversity functions at different levels and on it depends the stability of the biosphere, which in turn leads to the stability in climate, water, soil, chemistry of air and overall health of biosphere at large. It is the source of species on which human race depends for food, fodder, fuel, fibre, shelter and medicine etc. The countries located in the tropical and subtropical belts are far richer in biodiversity than the countries in the temperate region. Significant losses of biological diversity could affect the future of human life. Extinction of species diminishes future resource options. Man-made threats include destruction of habitat, over exploitation of natural resources, overgrazing, shifting cultivation, industrialization, urbanization, construction of dams and roads and mining etc. continuous degradation of environment, reflecting various interlinked problems such as over population, deforestation, poverty, pollution and global warming have lead the scientific community to realize that living systems, integrating the whole range of biological components are crucial for survival (Heywood, 1995). Only by conserving and regenerating our forests can we avert the dangers inherent in the loss of green cover involving them in the management of forests their indigenous knowledge of sustainable use of resources.

The present study has been undertaken in the Renuka forest division, which supports rich and diverse flora on one hand and where all the above mentioned biotic as well as abiotic interferences are actively taking place. The phenomenon leading to loss in the natural habitat which might prove to be hazardous to the rich floral diversity in future and as such needs serious scientific investigation to allow the forests to exist on sustainable bases.

1.2 STATEMENT OF THE PROBLEM

Forests provide many social, economic, and environmental benefits. In addition to timber and paper products, forests provide wildlife habitat and recreational opportunities, prevent soil erosion and flooding, help provide clean air and water, and contain tremendous biodiversity. Forests are also an important defence against global climate change. Through the process of photosynthesis, forests produce life-giving oxygen and consume huge amounts of carbon dioxide, the atmospheric chemical most responsible for global warming. By decreasing the amount of carbon dioxide in the atmosphere, forests may reduce the effects of global warming. The large tracts of the richest forests in the world have been cleared for wood fuel, timber products, agriculture, and livestock. These forests are rapidly disappearing. Clearing of forests was started by man around 8000 B.C. when the cultivation of crops was started. Since then about 50 percent of the forests that covered the earth have been converted to farms, pastures, and other uses. But, the human impact on forests did not stop there. Most of the forests that are left have been heavily altered by humans, often rendered into a patchwork of smaller forested areas (Mitchell, 1989).

According to the National Forest Policy 1988, the minimum desired area, which is considered safer for a tropical country like India is about 33 percent. As per broad policy recommendations, about 66 percent of the area in the Himalayas and the Peninsular hills and 25 percent in the Great Plains should be under forests (Anon, 1988). These figures accentuates that there is deficiency of forest cover across the country due to many factors such as illicit felling of tree, expansion of shifting agriculture and illegal encroachment, over grazing, wrong practices of farming, excess taping of resin, unscientific quarrying etc. Forests are most important entity to the livelihoods of the majority people in Himachal Pradesh. Rural people comprise around 90 percent of Himachal Pradesh's population, and most of them are dependent on forests to some extent (e.g. for fuel wood, fodder, grazing, construction timber, non-timber forest products). Currently forest resources tend to be used by local communities for subsistence needs, subject to restrictions. The HP Forestry Department (HPFD) has formal control of forest use and draws revenue from timber and resin production (Forest Statistics, 1998).

However, there is also a high level of illicit extraction of produce often to distant markets. Rural people feel excluded from control of forests and, with no legitimate local organisational basis for resource use regulation, extract products and graze livestock without effective self-regulation. The outcome, where HP forestry department regulation is also weak, is resource degradation, particularly of grazing lands, and un-demarcated protected forests. There is great potential for local resource management through participatory use regulation, and in some places there is potential for building forest-based enterprises in production of forest goods and tourism for example.

In Renuka forest division, of Sirmour district, the threats include encroachments, forest fires, illicit felling for timber and grazing, diseases and incursion of weeds and other invasive species. The study area is rich in mineral resources, the important ones being marble, limestone, slate and different types of building materials, whose exploitation is being done on a large. The Kafota and Sangrah ranges have large patches of mining areas being practised and it resulted into loss of valuable forests bearing trees and other forms of vegetation. The resin extraction has also proved a major cause of forests depletion at the large scale in the study area.

The fire is common in the Chil (*Pinus roxburghii*) forests in the study area. The main cause of fire has been human negligence and other reason of fire is the careless burning of the pastures by locals wherein the fire escape into the forests. Yet another problem is the excessive use of acid during resin tapping harms the Chil trees and these turn dry. The Oak is the main target of lopping. It is mainly lopped for fodder and fuel. Deodar is also lopped for fuel and manuring. Spruce is mainly lopped by the Gujjars for the roofing of their huts. Grazing has a dual impact on the forest. First the browsing and trampling of the seedlings of plantations and natural regeneration are destroyed and secondly, the soil at the paraos (camps) is trampled by buffaloes, sheep and goats and this leads to erosion.

The roads which are considered the life lines of progress and development of any area has been detrimental to the forest area. The cutting of hills sides for roads and buildings results in the sliding of the land on one hand while the dumping of the debris downhill results in damaging of the forests down below the slopes. The Renuka and Kafota ranges have large deposits of limestones. The mining activities in both the private and government lands are proving detrimental as no efforts to reclaim these areas have been undertaken. The Lantana camara and Euphorbia royleana are two plants which are invading the forest area. Illegal encroachment is also a problem which causes the damage to the forest depletion in the study area.

1.3 REVIEW OF LITERATURE

The climate plays a predominant role in the formation of bio-environment and it decides the vegetation and animal associations. Various social scientist and anthropologists have been involved for many years in the study to the process of evolution and origin of different societies in the world **(Sharafeldin, 1982)**. A lot of work has been done in this direction that perhaps falls beyond the scope of present study. However, the study of forest ecosystems and other related aspects concerning particularly to geographers are vital to the present study. Such studies wherever conducted will guide and form the basis for the present study. In this context it is important to say that not much work has been conducted particularly by geographers pertaining to forests of Himachal Pradesh. Thus it can be said that as far as the forest resources in the state of Himachal Pradesh in general and Sirmour district, in particular and especially in Renuka forest division are concerned, they have not been studied and no systematic work from ecological, botanical and even geographical point of view has been attempted. However, the research work done in the field of forests in different parts of the world, has been analyzed under the six themes.

1.3.1 FOREST STUDIES IN GENERAL NATURE

Gaston et al. (1981) opined that the maximum species diversity is in the rain forest; however, Himalayan environment is more diverse at levels of biological organization, higher than species, genera, family, habitats and ecosystems. It is further stated that no other life zone contains so much variation between habitat and ecosystem as mountains. It has striking zonation of natural vegetation besides occurrence of endemic species only with narrow range. The narrow range in the mountains according to Gaston et al. is more pronounced because of micro-variation in climate, soil and vegetation over very short distances. **(Gadgil, 1994; Gadgil and Guha, 1995; Heywood, 1995)** believed that human influence on ecosystem is exponentially increasing not solely due to population explosion, but partly because of ever accelerating resource demand by the so called, biosphere people, that mainly comprise the third world elite and citizens of the industrial countries. **(IUCN, 1994)** expressed that these interventions coupled with certain natural calamities such as frequent earthquakes, cyclones and natural fires etc. have given way to rapid genetic erosion with extinction rate as high as 30 percent per decade of global species.

Mahajan (1983) studied that utilization is the basic cause of depletion of living resource all over the world and Himalayas is no exception to it. Under these circumstances we are left with no other option except to develop a judicious management practices for the rational utilization particularly of the renewable resources. **Lal (1989)** stated that species diversity can be easily measured, while association diversity may not be and there are two main reasons for the same. Firstly, associations are not discrete units, which can be easily defined on the ground. Secondly there is no universally accepted operational definition of biological diversity. **Fernandez et al. (1996)** gave brief description of the physiography, geology, climate and vegetation types of the Balas river basin in central-south Mexico. The floristic list includes 202 families, 1246 genera and 4442 species. Forest vegetation was investigated by species diversity and species sequence importance analysis by **Hwan et al. (1999)**. Vegetation was classified into 10 plant communities. Five measures of species diversity and their relationship with altitude, soil factors (pH, base content) and community type were studied. A perusal of literature on the ecology of the Himalayan vegetation reveals that the most of the phyto-sociological and other quantitative studies have been conducted in Kumaun and Garhwal Himalaya **(Gupta, 1972; Singh and Singh 1987 and 1992)**. The most significant works include classification of forest formations **(Champion and Seth 1968; Schweinfurth, 1968)**. Variuos studies include the biomass productivity of various communities seasonality and productivity of alpine vegetation **(Ralhan et al. 1984)**.

Cheng et al. (1998) studied that major causes of deforestation and degradation of natural resources in the Belete-Gera forest of Ethiopia are coffee production activities and encroachment into forestland to expand farmland and pasture. Population growth, and the government's land reform and resettlement programs have caused local residents to lose harmony with the land. Forest management in this area hasn't yet been fully developed. The objectives of this study are to identify the extent of deforestation and natural resource degradation, in preparation for a sound management plan. Encroachment of farmland and pasture into natural forest during the past four years has been identified through interviews and aerial photo interpretation. The encroachment rate is 1.45 percent per year. Encroachment occurred mostly on areas with gentle slopes adjacent to populated villages and along roads and footpaths. The extent and impact of coffee production activities were examined through agency documents, forest survey data and vegetation survey. It is estimated that up to 49 percent of the accessible natural forest is under the influence of coffee production activities, among which collecting of naturally grown coffee beans has the least and the coffee plantations has the most impact on the natural forest. Coffee plantations in natural forest have reduced the forest density and species diversity. Age structure of the trees is limited to mature and old classes only, which eventually endangers their function as shade for coffee plantations.

Dongsheng et al. (2004) indicated that China is rich in forest resources and diversified environments, covering a vast territory. The changes in forest resources have a direct bearing on environmental quality. The study gives a detailed account of the dynamic change in forest resources in China, including the overall process of forest evolution, the status quo, and features of the existing forest resources and the development and use of major forest resources. In addition, it analyses the current situation of China's environment and explores the main contributing factors based on the overall environmental situation. To achieve sustainable management of forest resources and improve the environment in China, the Chinese Government attaches great importance to the protection and development of forest resources as well as to environmental development and improvement. The study gives an overview of current thinking for sustainable forest resource and environmental development in the future and highlights the current focus of efforts toward this objective.

Birendra and Shin (2006) analyzed the relationship between forest resources, refugees, and the host population. The findings of the research suggest that the host populations are heavily dependent on the local forest for their daily needs such as fuelwood, timber, grazing area, fodder for domestic animals, foods, and medicine in addition to cultural and esthetic needs. The forest has also been relied upon for agricultural needs such as manufacture of agricultural tools, maintenance of irrigation water systems, erosion control, and fertilizer needs. The forest was under a sustained demand as any other Terai forest of Nepal. After the arrival of refugees in 1992, the demand for forest resources increased substantially. Initially, the construction of the refugee camps decreased the total forest area and also required some felling of trees. More significantly, the refugees themselves became active users of the forest resource, which generated extra pressure on the forest and created scarcity of forest resources. Before the arrival of the refugees, forest management and monitoring of illegal use of the forest resources were carried out by the government through its local forester office. The local residents were active users of the forest resources, but were passive in managing and maintaining the forest resource.

Ota et al. (2011) evaluated the influence of texture information from remote sensed data on the accuracy of forest type classification at different spatial resolutions. And used 4 m spatial resolution imagery to create five different sets of imagery with lower spatial resolutions down to 30 m. They classified forest type using spectral information alone, texture information alone, and spectral and texture information combined at each spatial resolution, and compared the classification accuracy at each resolution. The classification and regression tree method was used. The accuracy of all three tests decreased slightly with lower spatial resolution. The accuracy with the combined data was generally higher than with either the spectral or texture information alone. At most resolutions, the lowest accuracy was with texture information alone. However, there was no clear difference in accuracy between the combined data and spectral data alone at 25 m and 30 m spatial resolution. These results indicate that adding texture information to spatial information improves the accuracy of forest type classification from very high resolution (4 m spatial resolution) to medium resolution imagery (20 m spatial resolution), but this accuracy improvement does not appear to hold for relatively coarse resolution imagery (25 m to 30 m spatial resolution).

Miah et al. (2012) stated that the natural forests offer numerous benefits to indigenous communities and society at large. Incomes from forest sources play an important role in rural households. In addition to this, environmental sources in the forests contribute significantly to rural households' livelihoods and economic well-being. The study examines the contributions of forests to the livelihoods of the Chakma tribe in Bangladesh. Using the data from 60 randomly sampled households from three villages, it measured forest-resource use with a monetary yardstick. As revealed through analyses, natural forest-sourced income occupies the second-largest share in total average household income next to shifting cultivation income in the study area. It was also observed that larger families with more people gathering forest products realized more forest income. This study will be relevant to forest and environmental policy-makers as well as indigenous community development practitioners.

Phung et al. (2014) formulated documents how the implementation of forest tenure policy affects the decision-making of farmers in mangrove-shrimp farming systems with regard to their access to and management of mangrove forest in Ca Mau, Mekong Delta, which is the largest remaining mangrove forest in Vietnam. Policies on land allocation, land tenure and use-rights are important since they potentially promote sustainable mangrove-shrimp management. Forest management policy in Vietnam has been changed to promote equality of benefit sharing among stakeholders and devolved State forest management to the household level. However, to what extent its implementation can stimulate both mangrove conservation and livelihood improvement is still being debated. They tried to access its social mechanisms to investigate how state forest companies (FC) and farmers can benefit from mangrove exploitation. The study was conducted from September 2008 to August 2010 using both qualitative and quantitative methods and using a participatory approach. After group discussions and in-depth interviews with a wide range of stakeholders, they interviewed 86 households in four communities using structured questionnaires. Results show the imbalance in access to finance, markets, and differences in authority between the two actors, farmers and FC. The discussion focuses on the possibilities of win-win outcomes, i.e. land tenure regimes promoting the devolution of sustainable forest management to farm households to balance benefits of both mangrove conservation and livelihood improvement.

1.3.2 PHYTOSOCIOLOGY

The plant sociology (or phytosociology) is defined as the discipline which concerns itself with the study of vegetation as such, with its floristic composition, structure, development and distribution, whereas the term ecology is restricted to the study of the habitat **(Tansley, 1920)**. This divergence in aim and difference in origin has led to a lack of sympathy between the two. The phytosociologists have naturally become concerned with producing a complete description and classification of vegetation, while the British ecologists, once the early enthusiasm for primary survey had diminished. **(Tansley, 1947)**, devoted themselves to the investigation of the physiological relationship between plant and habitat, with the study of succession and development, and with the perfection of methods for the objective description of the plant community as a basis for more exact synecological studies. Points of contact between these two schools of thought have become fewer because their aims are nearly mutually exclusive; intensive work on ecological problems and in the field of experimental taxonomy have raised doubts on the plant species growing together have mutual relation among themselves and also with the surroundings **(Lugo, 1978)**. From habitat to habitat population of plant communities varies and this helps in finding out the type of plant community. **Walter (1979)** holds the widespread assumption that plant species distribution is directly dependent upon the physical conditions prevailing in a habitat is incorrect, as these conditions are of importance only indirectly. A complete list of the species present on each plot is considered an almost absolute requirement for the working out of the sociology of the growing species. The studies of phytosociology help a deeper insight into the structure composition of the vegetation. The history of the application of sigmatist phytosociology to nature conservation and land use planning in Spain in quite brief. The use of the historical information including the formal documented records and the non-formal indigenous knowledge can provide vital information on indigenous resource management system and the present day management methods. There is a need to develop sampling methods and protocol that allow reliable comparison between sites without a complete inventory being taken. **Schnitzler (1994)** studied conservation of biodiversity in alluvial hardwood forests of the Temperate Zone. He observed that the distribution of wetland temperate forests is high during past three decades throughout the world. He gave an overview of the ecological features of Rhine forests in Alsoce (France) and Baden-Württemberg and Hesse (Germany) both before and after human alteration and proposes a planning strategy for their conservation. **Chytry (2001)** studied the large phytosociological data sets of three types of grassland and three types of forest vegetation from the Czech Republic were analysed with a focus on plot size used in phytosociological sampling and on the species-area relationship. The data sets included 12975 releves, sampled by different authors in different parts of the country between 1922 and 1999. It was shown that in the grassland data sets, the releves sampled before the 1960s tended to have a larger plot size than the releves made later on. No temporal variation in plot sizes used was detected in forest releves. Species-area curves fitted to the data showed unnatural shapes, with levelling-off or even decrease in plot sizes higher than average. This distortion is explained by the subjective, preferential method of field sampling used in phytosociology. When making releves in species-poor vegetation, researchers probably tend to use larger plots in order to include more species. The reason for this may be that a higher number of species gives a higher probability of including presumed diagnostic species, so that the releves can be more easily classified in the Braun-Blanquet classification system. This attitude of phytosociologists has at least two consequences: (1) in phytosociological data bases species-poor vegetation types are underrepresented or releves are artificially biased towards higher species richness; (2) the suitability of phytosociological data for species richness estimation is severely limited. **Shackleton and Pandey (2014)** studied that non-timber forests products (NTFPs) provide multiple livelihood benefits to local communities and regional and national economies. And yet this knowledge is rarely drawn upon in debates around and design of poverty alleviation or land use policies, strategies and projects. Unless the accumulating wealth of empirical evidence can be translated into policies and approaches at higher levels, and integrated into poverty alleviation programmes, it will have little impact on local and national poverty profiles. In this study they discussed eight steps to facilitate integration of NTFPs into the development agenda, for the benefit of local communities.

1.3.3 FLORISTIC COMPOSITION/DIVERSITY

Dansereu (1960) considered floristic composition as one of the major distinguishing characters of the community because each of the species of a community has not only its own ecological amplitude, but also its particular relationship to the environment and its associated species. Thus, the nature of a plant community at a place is determined by the species that grow and develop in such environment. **Bliss (1966)** studied the climax vegetation of a monsoon deciduous forest with high species diversity and stated that clearing the forest rapidly gives rise to savanna, which on regrowth degenerates into grasslands with continuous grazing and further felling. **Gentry (1988)** studied the trends in community composition and diversity of neotropical forests as measured by a series of samples of and analyzed as a function of various environmental parameters. These trends have also compared with those found in similar data sets from other continents. Altogether the basic 0.1-ha data sets are reported for 87 sites in 25 countries on six continents and several islands. New data from ten 1-ha tree plots in upper Amazonia are also compared with each other and with similar data from the literature.

Solangaarachchhi and Perera (1993) reported details of floristic information on 81 species of important medicinal plants representing 73 genera and 37 families in the understorey of the tropical dry mixed evergreen forest at Hurula reserve of Sri Lanka. About 1 percent of the species were endemic and 61 percent were found to be of medicinal importance. While studying species composition, diversity index and regeneration potential of forest communities of outer Garhwal, **Ram Prasad et al. (1992)** reported that lower altitudinal forests (700-900 m) were dominated by *Shorea robusta/Mallotus philippensis* and *Shorea robusta/Pinus roxburghii* and mid altitudes (1000-1700 m) by *Pinus roxburghii* and those of greater than 2000 m altitude by *Quercus incana/ Rhododendron arboreum*. Maximum similarity in species composition was found in *Pinus/Quercus* and *Quercus/Rhododentron* communities and maximum diversity in *Shorea/Pinus* and *Quercus/Rhodendron* communities. **Jain et al. (1994)** reported that more than 600 plant species grow around the Sahariya-inhabited area of Central India which is used for different purposes like fuel, food, fibre, medicines, small wood timber etc. A number of clans are named after plants, which they never harm.

Gupta, Chaudhary and Wate (2008) carried out to assess the diverse floristic wealth in urban forest area of NEERI campus at Nagpur, Maharashtra (India). This urban forest is ecologically important to maintain the atmospheric temperature around 2 degrees C below and higher relative humidity as compared to other urban areas. The water table is also observed to be shallower in this area as compared to other areas. Therefore, the biological diversity of this urban forest was studied, as it is directly related to ecology of the area. Floristic survey of NEERI premises recorded 135 vascular plants including 16 monocots and 119 dicots, belonging to 115 genera and 53 families. The taxa included 4 types of grasses, 55 herbs, 30 shrubs and 46 trees. The large number of species within very small area (43 ha) indicates rich biodiversity in this forest area. It is also observed that this forest patch has tall trees, with good density and rich cover of shrubs and herbs on forest floor indicating well knit plant community. These characteristics have given immense ecological importance to this urban forest area. Detailed vegetation study revealed that positive co-operation in the plant communities can significantly maintain species diversity in the environment.

Ellum (2009) studied that current interest in the conservation of biodiversity is generating a need for forest management and silvicultural techniques designed to maintain the integrity of ecosystems while satisfying society's need for timber resources. The conservation of forest understory plant communities should be a major emphasis of this effort as they contain the majority of plant diversity in most U.S. forests and play a significant role in many ecosystem functions. This study reviews the literature regarding plant responses to disturbance-most importantly changes in light environments-and applies that information to forest management. A comparison of developmental plasticity and rapid acclimation as response pathways is used to discuss plant level responses. At the landscape level, diversity models, silviculture treatments, and site characteristics are used to discuss changes in understory community composition following disturbance. Results of ongoing research on the effects of forest management on floristic patterns for southern New England forests are summarized.

Chun (2014) stated that Indigenous people have a vital role in environmental management and development because of their knowledge and traditional practices. The concept of sustainable development requires taking consideration into the legal system of traditional knowledge for the benefit of indigenous people who live in or near the forest. In modern administrative countries, the statutory legal system established by the government had almost prevailed over the common law system especially relating to management of natural resources. The statute regulating forest resources should prescribe the inherent interests to traditional knowledge of indigenous people in the forest. To endow indigenous people with inherent interests about forest resources and rights to traditional knowledge, the statute legal system and the common law system have to cooperate with each other according to the governance theory of cross-regulation.

1.3.4 FOREST STUDIES IN HIMACHAL PRADESH

Floristic composition of sub-alpine Western Himalayan coniferous forest has been found to be dominating by *Abies spectabilis* in the Parbati Range in Himachal

Pradesh **(Gupta et al. 1982)**. Detailed information regarding scientific, vernacular names, distribution pattern, official parts and uses of 55 medicinal plants used in traditional medicinal systems and collected on commercial scale in Kinnaur region **(Singh, 1990)**. **Nautiyal et al. (1994)** conducted exploration of fodder grasses and legumes germplasm of cold desert of India viz., Ladakh region in Jammu and Kashmir and Lahaul-Spiti in Himachal Pradesh. A total of 61 accessions of 16 fodder legumes and 160 accessions of 63 fodder grasses were documented for conservation and multiplication along with 21 woody fodder plant species. In the same year, Aswal and Mehrotra, did phytogeographical studied of the flora of Lahaul-Spiti, revealed that the flora showed more affinity with the mountain flora of the Middle Asia, although a large number of plants are common to north temperate and Arctic region of the Eurasia, the Mediterranean region, the Chinese Mountains and the higher plateau of Tibet. In their book "Flora of Lahaul-Spiti" comparison of the ten dominating families of the angiosperms in the Lahaul-Spiti; Western Himalayan, Eastern Himalaya, and British India, it was indicated that Asteraceae which occupied first position in the Lahaul-Spiti, is second, fourth and seventh in Western Himalayas, the Eastern Himalayas and other parts of India respectively. The flora of the region showed close affinity with the high altitude flora of Western Himalayas **(Aswal and Mehrotra, 1994)**.

 Joshi et al. (2001) in their study they used wide resolution satellite data for tropical forest spectral discrimination and mapping at a regional scale. Himachal Pradesh was selected as a case study using a multitemporal WiFS data set of 1998. This study documents first the relevance of WiFS data to assess the extent of seasonal forest. A phenologically dependent methodology is developed for the Himalayas, where generally snow/cloud-free data set are hardly available. Unsupervised classification using maximum normalized difference vegetation index, band 1 and band 2 data was processed. The hybrid approach was used to refine the classes. The classes were labelled using the spectral values from ground truth, available data sets and spectral analysis of the data sets. For evaluation of classification, comparisons were made at a regional level with the available forest database. The classes and statistics were in correlation to the ground reality. The estimated forest was 17.15 percent, whereas the forest cover reported by the Forest Survey of India is 22.5 percent of the total geographical area of the state.

Kumar et al. (2012) studied the ethno-medicinal plants at Kanag Hill, Tehsil Theog in Shimla district. The study was mainly focused on medicinal plants used for treatment of various ailments by the nearby village inhabitants. The information was collected by questionnaire and consulting local old people. The study entirely focused on revealing the medicinal potential possessed by the plants growing wild in this area and their sustainability for the betterment of mankind. Kumar and Choyal (2013) conducted, an ethnobotanical survey of the plant diversity in lower foothills villages of six districts i.e. Kangra, Hamirpur, Mandi, Una, and Bilaspur of Himachal Pradesh. The study was mainly focused on the traditional uses of the 32 medicinal plants of lower foothills used for the treatment of oral health problems and other mouth disorders of nearby village inhabitants. The information was carried out by the personal interviews of local old people.

1.3.5 FOREST COVER CHANGE, REMOTE SENSING AND GEOGRAPHIC INFORMATION SYSTEM

Forest is a complex ecological system in which trees are dominant life forms. The word 'Forest' is derived from Latin word 'Foris' meaning outside, the reference apparently being to a village, boundary or fence. Thus, originally, a forest must have included all uncultivated and uninhabited land. Today, a forest is any land managed for the diverse purposes of forestry, whether or not covered with trees, shrubs, climbers or such other vegetation. Forest is striking feature of the land system. The forests of a country are natural assets of great value, which represents largest most complexes and most self-generating of all ecosystems. They covered about one-third of the land area of the world and constitute one-half of the total biomass. Forest cover is the area covered under vegetation with a tree canopy cover more than 10 percent (FAO, 2000). Human activity is vastly altering the Earth's vegetation cover. Such changes have considerable consequence for the health and resilience of ecosystem and contribute to anthropogenic climate change through a variety if processes. These include the growth or degradation of surface vegetation which produces changes in the global atmospheric concentration of carbon dioxide; and changes in the land surface, which affect regional and global climate by producing changes in the surface energy budgets (Vitousek, 1994). There have been many change detection studies as part of environmental impact assessment programmes in India and abroad. Here some Indian examples are cited. (NRSA, 1983b, now NRSC) studied the land use/forest cover changes in Idukki Hydroelectric Project area. Two data sets pertaining to the periods 1978 and 1982 were used for this purpose. Post classification comparison technique of change detection was employed and changes in spatial extent in land use in general and forest cover in particular, were found out.

Kushwaha (1985) studied the long-term and short-term changes in the land cover on Sriharikota Island, India using 1973, 1983 and 1985 period Landsat MSS data and post classification comparison technique. The author observed that the forest vegetation on the Island had undergone tremendous changes since 1973 due to the combined impact of Indian Space Research Organization's activities and that of devastating November, 1984 Several deforestation studies based on samples of high-resolution imagery have been performed at regional scales **(FAO, 1996; FAO, 2000; Achard et al. 2002)**, but have lacked a precise method for targeting likely areas of change. Some researchers maintain that sampling high resolution imagery cannot efficiently estimate deforestation rates. The rarity of land cover types and changes, and the presence of outliers in the population, are prime culprits contributing to poor precision of sampling-based estimates of land cover composition and land cover change **(Stehman et al. 2003)**. This has reinforced the use of regional wall-to-wall mapping strategies **(Skole and Tucker, 1993; Townshend et al. 1995)**. The forest cover change in a shifting cultivation area in north-eastern India employing Image differencing methods and found that image differencing techniques were very effective in the assessment of positive and negative changes in forest vegetation in the study area **(Singh, 1989)**. Of all the available change detection methods, image differencing has been widely used. This method highlights the change areas more noticeably than any other change detection method. Satellite remote sensing has played a pivotal role in generating information about forest cover, vegetation type and land use changes **(Roy, 1993)**. The increase in processing speed and the compression techniques for digital storage have made digital imaging available to anyone. One advantage of digital imagery for natural resource managers is that it can be enhanced on the computer to bring out details of interest- whether vegetation stress, species composition, or growth and volume. The standardization of ground sampling methods, understanding of spectral and temporal responses of vegetation have brought acceptance of the application of satellite remote sensing data in forest inventory and mapping **(Roy, Dutt and Joshi, 2000)**. In the recent years, since 1972, there has always been at least one multispectral satellite with medium spatial resolution in operation and frequently more than one. Landsat, SPOT, IRS, MOS and IKONOS are among the satellite in this class. These data sets are potential enough to provide the information in 1: 250,000 scale. It is assessed to provide information in multilevel to the forest managers for rapid forest cover mapping, detailed forest cover mapping in divisional level and enable monitoring of rapid changes in the forests due to forest fire, shifting cultivation etc. The merged data set (Multispectral-High resolution) provides better feasibility and enhanced capability to evolve new programmes in the forest management sector, especially in the forest inventories and micro-planning like Joint Forest Management (JFM) activities. The detailed information at compartment and village level scales the need to protect forest and call for the involvement of forest dwellers in forest management.

Shosheng and Kutiel (1994) investigated the advantages of remote sensing techniques in relation to field surveys in providing a regional description of vegetation cover. The results of their research were used to produce four vegetation cover maps that provided new information on spatial and temporal distributions of vegetation in this area and allowed regional quantitative assessment of the vegetation cover. Many researchers have used remotely sensed imagery to monitor land cover changes through time, for example **(Mas, 1999)**. The use of satellite imagery allows researchers to inventory and study the state of vegetation in a large region, while reducing the need to be in the field. The results produced by remotely sensed imagery can provide valuable information to resource managers by aiding in the management decision-making process **(Franklin et al. 2002)**. Forest managers are frequently interested in information regarding canopy changes caused by short-term natural phenomena, including insects, flood, drought, human activities, and reforestation **(Coppin and Bauer, 1994)**.

The world's forests are changing in quantity and quality, and in both positive and negative ways **(FAO, 1992)**, this process is associated with social, economic and environmental factors. The conversion of forest covers in general has severe long term environmental and socio-economic consequences globally as well as locally such as global climate change, habitat fragmentation and degradation, species extinction **(Phong, 2004)**. **Woodcock et al. (2001)** produced a map of forest change in the Cascade Range of Oregon by applying a new approach to monitor large areas by extending the application of a trained image classifier to data beyond its original temporal, spatial, and sensor domains. The method showed accuracies comparable to a map produced with current state-of-the-art methods and their results highlighted the value of the existing Landsat archive and the importance for continuity in the Landsat program. **Kumar, Arockiasamy and Britto (2002)** studied forest type maps of 1990 and 1999 and the change detection between these periods and portrayed the status of different forest types and their rate of change/degradation. Among the five forest types semi evergreen and deciduous forest types experience considerably high pressure from the local people. The deciduous forest is in its early succession stage needs severe protection for further growth and establishment. These pressures not only results in the degradation of these forests but also affects the regeneration potential. Though vegetation mapping in any region on large scale could provide spatial information on forest cover and type it is necessary to conduct the phyto-sociological study in different forest types to understand the diversity and distribution of species, which are very much important for planning conservation strategies.

Desclee et al. (2006) developed a new method of change detection for identifying forestland cover change in temperate forests using three SPOT-HRV images covering a 10 year period. Combining the advantages of image segmentation, image differencing and stochastic analysis of the multispectral signal, this method is called object-based and statistically driven. **Kennedy et al. (2007)** described and tested a new conceptual approach to change detection of forests using a dense temporal stack of Landsat Thematic Mapper (TM) imagery. They used Trajectory-based change detection for automated characterization of forest disturbance dynamics. The authors proposed that many phenomena associated with changes in land cover have distinctive temporal progressions both before and after the change event, and these lead to characteristic temporal signatures in spectral space, so instead of searching for single change events between two dates of imagery, they searched for these idealized signatures in the entire temporal trajectory of spectral values. The authors pointed that, this trajectory-based change detection is automated and has the advantages of requires no screening of non-forest area, and requires no metric-specific threshold development. More advantage is that, the method simultaneously provides estimates of discontinuous phenomena (disturbance date and intensity) as well as continuous phenomena (post-disturbance regeneration). **Masek et al. (2008)** mapped the disturbance and early recovery of a forest across North America for the period 1990-2000 using the Landsat satellite archive. The detection was performed using the temporal change in a Tasseled-Cap (Disturbance Index) calculated from the two images. The authors used a sample of biennial Landsat time series from 23 locations across the United States for validation. Their results indicate disturbance rates of up to 2-3 percent per year across the US and Canada due to harvest and forest fire.

Pradhan and Awang (2008) presented capability of remote sensing and Geographical Information Systems (GIS) to evaluate forest fire susceptibility. Forest Fire locations were identified in the study area from historical hotspots data from year 2000 to 2005 using AVHRR NOAA 12 and NOAA 16 satellite images. Various other supported data such as soil map, topographic data, and agro climate was collected and created using GIS. These data were constructed into a spatial database using GIS. The factors that influence to fire occurrence, such as fuel type and Normalized Differential Vegetation Index (NDVI) were extracted from classified Landsat-7 ETM plus imagery. Slope and aspect of topography, were derived from topographic database. Soil type was extracted from soil database and dry month code from agroclimate data. Forest fire susceptibility was analyzed using the forest fire occurrence factors by likelihood ratio method. The results of the analysis were verified using forest fire location data. The validation results show satisfactory agreement the susceptibility map and the existing data on forest fire location. The GIS was used to analyze the vast amount efficiently, and statistical programs were used to maintain the specificity and accuracy. The result can be used for early warning, fire suppression resources planning and allocation.

Mitchard et al. (2009) conducted a long-term (1982-2006) quantification of vegetation change in a forest-savanna boundary area in central Cameroon. They performed across-calibrated normalized difference vegetation index (NDVI) change detection method to compare three high-resolution images from 1986, 2000, and 2006. The result indicated that the largest changes were in the lower canopy cover classes whereas the higher canopy classes showed significant positive change. The authors attributed this due to a reduction in human pressure caused by urbanization, as rainfall did not alter significantly over the study period. They stated another possibility that increasing atmospheric CO_2 concentrations are altering the competitive balance between grasses and trees. The study proved that forest encroachment into savanna is an important process, occurring in forest–savanna boundary regions across tropical Africa.

Sakthive et al. (2010) precisely analyzed the trend of forest cover changes over the time span of 70 years, in the Kalrayan hills in India using satellite data acquired in 1931, 1971 and 2001. The forest cover, in the study area, during 1931 and 1971 were derived from the Survey of India top sheets of 1931 and 1971. The authors focused on the role of remote sensing and geographic information system in assessment of changes in forest cover. Their results clearly proved that forest cover has increased between 1931 and 1971 and decreased between 1971 and 2001. Yadav (2010) stated that it is undisputed that the forestry sector has one of the major use of remote sensing data. Several studies have been conducted successfully by foresters, researchers, scientists, NGO's, and others for vegetation cover, monitoring of degraded areas, loss of forest cover, forest fires, habitat mapping to mention a few. Several Forest Departments in the country have now established sophisticated GIS and RS facilities. Some states such as Karnataka, MP, WB, Sikkim, and Assam have started preparing Working Plans based on the new technology. The Forest Survey of India, Dehradun has taken a laudable lead in this matter. Already well tested methodologies of sampling, inventorization and interpretation are available with them. However, it is important that the Working Plan and GIS are interfaced properly.

Bharti et al. (2011) reported an upward shift of timberline vegetation by 300 m in Nanda Devi Biosphere Reserve (NDBR). The detection performed by **Panigrahy et al. (2010)**, detected timberline changes in the subalpine vegetation, using post classification comparison method. The authors managed to compare fir patches to see the changes in timberline vegetation. The author recommend use of Temporal Trajectory Analysis (TTA) for detecting changes due to different phenol-phases and suggest large scale mapping of major vegetation categories using remote sensing and ground truth verification so that to detect minor vegetation change more accurately.

Hansen (2013) stated that quantification of global forest change has been lacking despite the recognized importance of forest ecosystem services. In his study, Earth observation satellite data were used to map global forest loss (2.3 million square kilometers) and gain (0.8 million square kilometers) from 2000 to 2012 at a spatial resolution of 30 meters. The tropics were the only climate domain to exhibit a trend, with forest loss increasing by 2101 square kilometers per year. Brazil's well-documented reduction in deforestation was offset by increasing forest loss in Indonesia, Malaysia, Paraguay, Bolivia, Zambia, Angola, and elsewhere. Intensive forestry practiced within subtropical forests resulted in the highest rates of forest change globally. Boreal forest loss due largely to fire and forestry was second to that in the tropics in absolute and proportional terms. These results depict a globally consistent and locally relevant record of forest change.

Stibig et al. (2014) evaluated the extent and trends of forest cover in Southeast Asia for the periods 1990-2000 and 2000-2010 and provides an overview on the main causes of forest cover change. A systematic sample of 418 sites located at the one-degree geographical confluence points and covered with satellite imagery of 30 m resolution is used for the assessment. Techniques of image segmentation and automated classification are combined with visual satellite image interpretation and quality control, involving forestry experts from Southeast Asian countries. Complementing the quantitative results by indicative information on patterns and on processes of forest change obtained from the screening of satellite imagery and through expert consultation, respectively, confirms the conversion of forest to cash crops plantations as the main cause of forest loss in Southeast Asia. Logging and the replacement of natural forests by forest plantations are two further important change processes in the region.

1.3.6 LAND USE / LAND COVER AND GEOSPATIAL TECHNIQUES

The terms Land use and Land cover are not technically synonymous; there are different definitions of land cover and land use among the relevant scientists. Therefore, a brief explanation about these two terms is provided under this theme from the Encyclopedia of Earth. In general, the term land use and land cover change identifies all kinds of human modification of the Earth's surface. Land cover refers to the physical and biological cover over the surface of land, including water, vegetation; bare soil, and or artificial structures **(Ellis and Pontius, 2006)**. Land use has a complicated expression with different views compared with the term land cover. In fact, social scientists and land managers characterize this term more general to involve the social and economic purposes. Natural science researchers classify the term land use in different aspects of human activities upon lands such as farming, forestry and man-made constructions.

In 1985, the U.S Geological Survey carried out a research program to produce 1:250,000-scale land cover maps for Alaska using Landsat MSS data **(Fitzpatrick et al. 1987)**. The State of Maryland Health Resources Planning Commission also used Landsat TM data to create a land cover data set for inclusion in their Maryland Geographic Information (MAGI) database. All seven TM bands were used to produce a 21-class land cover map **(EOSAT, 1992)**. Also, in 1992, the Georgia Department of Natural Resources completed mapping the entire State of Georgia to identify and quantify wetlands and other land cover types using Landsat Thematic Mapper TM data **(ERDAS, 1992)**. The State of southern Carolina Lands Resources Conservation Commission developed a detailed land cover map composed of 19 classes from TM data **(EOSAT, 1994)**. This mapping effort employed multi-temporal imagery as well as multi-spectral data during classification.

Turner II et al. (1995) believed that land use involves both the manner in which the biophysical attributes of the land are manipulated and the intent underlying that manipulation-the purpose for which the land is used. **Lambin et al. (2001)** made differentiation between land cover (i.e. whatever can be observed such as grass, building) and land use (i.e. the actual use of land types such as grassland for livestock grazing, residential area). The pace, magnitude and spatial reach of human alteration of the Earth's land surface are unprecedented. The Land use/ Land cover pattern of a region is an outcome of natural and socio-economic factors and their utilization by men in time and space. Changes in land cover (biophysical attributes of the Earth's surface) and land use (human purpose or intent applied of these attributes) are among the most important **(Turner et al. 1990; Lambin et al. 1999)**. Land use/ Land cover changes are so pervasive that, when aggregated globally, they significantly affect the key aspect of Earth system functioning. They directly impact biotic diversity world-wide **(Sala et al. 2000)** contribute to local regional climate change. Land cover is defined as the observed biophysical cover on the earth's surface whereas land use as the arrangements, activities and inputs that people undertake on a certain land cover type **(FAO, 2000)**. Thus, land cover corresponds to the physical condition of the ground surface, e.g. forest, grassland, agriculture land etc. while land use reflects human activities such as the use of land for different purposes as industrial zones, residential zones, and agricultural fields. This definition establishes a direct link between land cover and the action of people in their environment, i.e. land use may lead to land cover change **(Phong, 2004)**.

Land use/ land cover is a hybrid category. Land use denotes the human employment of the land and is studied largely by social scientists. Land cover denotes the physical ad biotic character of the land surface and is studied largely by natural scientists. Connecting the two are proximate sources of change, human activities that directly alter the physical environment. These activities reflects human goal that are shaped by underlying social driving forces. Contemporary global environmental change is clearly unique. The human reshaping of the earth has reached a truly global scale, is unprecedented in its magnitude and rate, and increasingly involves significant impacts on the biogeochemical systems that sustain the biosphere **(Meyer and Turner, 1992)**. The land use/ land cover pattern of a region is an outcome of natural and socio- economic factors and their utilization by men in time and space. Land is becoming scarce resource due to immense agricultural and demographic pressure. Hence information on land use/ land cover and possibilities for their optimal use is essential for the selection, planning and implementation of land use schemes to meet the increasing demands or basic human needs and welfare **(Vitousek, 1994)**. Land cover changes take two forms, conversion form and category of land cover to another and modification of condition within a category. It is regarded as the single most important variables of global change affecting ecological systems with an impact on the environment that is at least as large as that associated with climate change. It is particularly related to the increase of population and intensive agriculture **(Awasthi et al. 2002)**.

Meyer (1995) studied that every parcel of the land on the earth's surface is unique in the cover it possesses. Land use and land cover are distinct yet closely linked characteristics of the earth's surface. The land use defined that grazing agriculture, urban development, logging and mining among many others. While land cover categories could be cropland, forest, wetland, pasture, roads, urban areas among many others. The term land cover originally referred to the kind and state of vegetation, such as forest or grass cover but it has broadened in subsequent usage to include other things such as human structures, soil types, biodiversity, surface and ground water. Forces other than anthropogenic can alter land cover. Natural events such as weather, flooding, fire, climate fluctuation, and ecosystem dynamics may also initiate modification upon land cover. Globally, land cover today is altered principally by direct human use, by agriculture and livestock's rising, forest harvesting and management and urban and suburban construction and development. There are also incidental impacts on land cover from other human activities such as forest and lake damaged by acid rain from fossil fuel combustion and crops near cities damaged by troposphere ozone resulting from automobile exhaust. Land use is one of the main factor through which human influences the environment. Land use changes or modification have important environmental consequences through their impacts on soil erosion, water quality micro- climate, methane and Carbon dioxide emission (**Awasthi, 2004**). A Land use and land cover change has become a central component in current strategies for managing natural resources and monitoring the environmental changes. Land cover altered by the human or natural alterations of land cover plays a major role in global scale pattern of climate change and biogeochemistry of the Earth's systems.

The importance of investigating land cover dynamics as a baseline requirement for sustainable management of natural resources has been highlighted by many researchers involved in global change studies (**Mertens et al. 2000; Read et al. 2002**). These scientists has argued that a more focused management intervention requires information on the rates and the impacts of land cover changes as well as the distribution of these changes in space and over time. Changes in land-use land cover have important consequences for natural resources through their impacts on soil and water quality, biodiversity and global climatic systems (**Awasthi et al. 2002**). The number of people dependent on agriculture is rising, mostly by encroaching upon forest area (**UNEP, 2001**). Land use affects land cover and changes in land cover affect land use. A change in either however is not necessarily the product of the other. Changes in land cover by land use do not necessarily imply degradation of the land. However, many shifting land use patterns driven by a variety of social causes, result in land cover changes that affects biodiversity, water and radiation budgets, trace gas emissions and other processes that come together to affect climate and biosphere (**Riebsame, Meyer and Turner, 1994**). Forces other than anthropogenic can alter Land cover. Natural events such as weather, flooding, fire, climate fluctuations, and ecosystem dynamics may also initiate modifications upon land cover. Globally, land cover today is altered principally by direct human use: by agriculture and livestock raising, forest harvesting and management and urban and suburban construction and development. There are also incidental impacts on land cover from other human activities such as forest and lakes damaged by acid rain from fossil fuel combustion and crops near cities damaged by tropospheric ozone resulting from automobile exhaust **(Meyer, 1995)**. Hence, in order to use land optimally, it is not only necessary to have the information on existing land use land cover but also the capability to monitor the dynamics of land use resulting out of both changing demands of increasing population and forces of nature acting to shape the landscape.

Conventional ground methods of land use mapping are labor intensive, time consuming and are done relatively infrequently. These maps soon become outdated with the passage of time, particularly in a rapid changing environment. In fact according to **Olorunfemi (1983)**, monitoring changes and time series analysis is quite difficult with traditional method of surveying. In recent years, satellite remote sensing techniques have been developed, which have proved to be of immense value for preparing accurate land use land cover maps and monitoring changes at regular intervals of time. In case of inaccessible region, this technique is perhaps the only method of obtaining the required data on a cost and time-effective basis. A remote sensing device records response, which is based on many characteristics of, the land surface, including natural and artificial cover. An interpreter uses the element of tone, texture, pattern, shape, size, shadow, site and association to derive information about land cover.

The generation of remotely sensed data/images by various types of sensor flown aboard different platforms at varying heights above the terrain and at different times of the day and the year does not lead to a simple classification system. It is often believed that no single classification could be used with all types of imagery and all scales. To date, the most successful attempt in developing a general purpose classification scheme compatible with remote sensing data has been by Anderson et al., which is also referred to as USGS classification scheme. Other classification schemes available for use with remotely sensed data are basically modification of the above classification scheme. Ever since the launch of the first remote sensing satellite (Landsat-1) in 1972, land use land cover studies were carried out on different scales for different users. For instance, waste land mapping of India was carried out on 1:1 million scales by NRSA using 1980-82 Landsat multi spectral scanner data. About 16.2 percent of waste lands were estimated based on the study. **Xiaomei and Ronqing (1999)** noted that information about change is necessary for updating land cover maps and the management of natural resources. The information may be obtained by visiting sites on the ground and or extracting it from remotely sensed data.

Change detection is the process of identifying differences in the state of an object or phenomenon by observing it at different times **(Singh, 1989)**. Change detection is an important process in monitoring and managing natural resources and urban development because it provides quantitative analysis of the spatial distribution of the population of interest. **Macleod and Congation, (1998)** listed four aspects of change detection, which are important when monitoring natural resources; detecting the changes that have occurred, identifying the nature of the change, measuring the area extent of the change   and assessing the spatial pattern of the change. The basis of using remote sensing data for change detection is that changes in land cover result in changes in radiance values, which can be remotely sensed. Techniques to perform change detection with satellite imagery have become numerous as a result of increasing versatility in manipulating digital data and increasing computer power. A wide variety of digital change detection techniques have been developed over the last two decades. **Singh (1989) and Coppin and Bauer (1996)** summarized eleven different change detection algorithms; delta or post classification comparisons, multidimensional temporal feature space analysis etc. An analysis of land use and land cover changes using the combination of MSS Landsat and land use map of Indonesia **(Dimyati, 1995)** reveals that land use land cover change were evaluated by using remote sensing to calculate the index of changes which was done by the superimposition of land use land cover images of 1972, 1984 and land use maps of 1990. This was done to analyze the pattern of change in the area, which was rather difficult with the traditional method of surveying as noted by **Olorunfemi (1983)** used aerial photographic approach to monitor urban land use in developing countries with Ilorin in Nigeria as the case study.

Adeniyi and Omojola, (1999) studied the land use land cover change evaluation in Sokoto-Rima Basin of North-Western Nigeria based on Archival

Remote Sensing and GIS techniques, used aerial photographs, Landsat MSS, SPOT XS/Panchromatic image transparency and topographic map sheets to study changes in the two dams (Sokoto and Guronyo) between 1962 and 1986. The work revealed that land use land cover of both areas was unchanged before the construction while settlement alone covered most part of the area. However, during the post - dam era, land use /land cover classes changed but with settlement still remaining the largest. In some instance, land use land cover change may result in environmental, social and economic impacts of greater damage than benefit to the area **(Moshen, 1999)**. Therefore, data on land use change are of great importance to planners in monitoring the consequences of land use change on the area. Such data are of value of resources management and agencies that plan and assess land use patterns and in modeling and predicting future changes. Neither population nor poverty alone constitutes the sole and major underlying causes of land cover change worldwide. Rather people's response to economic opportunities as mediated by institutional factors drive land covers changes **(Lambin et al. 2001)**. It has been noted over time through series of studies that Landsat Thematic Mapper is adequate for general extensive synoptic coverage of large areas. As a result, this reduces the need for expensive and time-consuming ground surveys conducted for validation of data. Generally, satellite imagery is able to provide more frequent data collection on a regular basis unlike aerial photographs which although may provide more geometrically accurate maps, is limited in respect to its extent of coverage and expensive; which means, it is not often used.

Daniel et al. (2002) conducted comparison of land use land cover change detection methods, made use of five methods viz; traditional post classification cross tabulation, cross correlation analysis, neural networks, knowledge based expert systems, and image segmentation and object -oriented classification. Nine land use land cover classes were selected for analysis. They observed that there are merits to each of the five methods examined, and that, at the point of their research; no single approach can solve the land use change detection problem. **Pandey et al. (2006)** carried out a study on land use land cover mapping Panchkula, Ambala and Yamunanagar districts of Haryana state in India. They observed that the heterogeneous climate and physiographic conditions in these districts has resulted in the development of different land use land cover in these districts, an evaluation by digital analysis of satellite data indicates that majority of areas in these districts are used for agricultural purposes. The hilly regions exhibit fair development of reserved forests. It is inferred that land use land cover pattern in the area are generally controlled by agro- climatic conditions, ground water potential and a host of other factors. **Opeyemi (2006)** studied land use land cover changes in Ilorin between 1972 and 2001 and also an attempt was made at projecting the observed land use land cover in the next 14 yrs. Land consumption rate and land absorption co-efficient were introduced to aid in the qualitative assessment of the change. The results of the work show a rapid growth in built-up land between 1972 and 1986 while the periods between 1986 and 2001 witnessed a reduction in this class.

Bhattrai and Dennis (2007) studied the land use dynamics and forest cover change in Nepal's Bara District (1973-2003). Rapid land use and cover dynamics was seen near Simara, Amlekhjunga, Pathlaiya, Dumarwana and Nijgadh, suggest that there has been attempt to convert hitherto non-agriculture land to agriculture land. Such farm plantation has probably occurred in response to the diversification of household occupations from on-farm to off- farm. These observations support one of the propositions of ecological modernization theory that people respond to environmental degradation by changing their socio-economic activities. Network in bank landscapes (NBL), one of the Biodiversity hotspots, has undergone modification due to the anthropogenic factors **(Sharma et al. 2010)**. Excessive encroachment has also resulted in conversion of forest area into agriculture land. Land cover is a basic parameter, which evaluates the content of earth surface as an important factor that affects the condition and functioning of the ecosystem. Land cover is a biophysical state of the Earth surface, which can be used to estimate the interaction of biodiversity with the surrounding environment. Nowadays, land use land cover analysis plays an important role in the field of environmental science and natural resource management. The Land cover reflects the biophysical of state of the earth's surface and immediate surface, including the soil material, vegetation and water. Land use refers to utilization of land resources by human beings and land cover changes often reflects the most significant impact on environment due to excessive human activities. Land use and land cover is dynamic in nature and provides a comprehensive understanding of the interaction and relationship of anthropogenic activities with the environment **(Prakasam, 2010)**. Land use/cover changes also involve the modification, either direct or indirect, of a natural habitats and their impact on the ecology of the area. Land use/cover change has become a central component in current strategies for managing natural resource and monitoring environmental changes **(Tiwari and Saxena, 2011)**. Land use/cover pattern of a region gives information about the natural and socio-economic factors, human livelihood and development.

Song et al. (2011) stated that there is a need for improved and up-to-date land use/land cover (LULC) data sets over an intensively changing area in the Amur River basin (ARB) in support of science and policy applications focused on understanding of the role and response of the LULC to environmental change issues. The main goal of this study was to map LULC in the ARB using MODIS 250-m Normalized Difference Vegetation Index (NDVI), Land Surface Vegetation Index (LSWI), and reflectance time series data for 2001 and 2007. Another goal was to test the consistency of the classification results using relatively coarse resolution MODIS imagery data in order to develop a methodology for rapid production of an up-to-date LULC data set. The results on MODIS land cover were evaluated using existing land use/cover data as derived from Landsat TM data. It was found that the MODIS 250m NDVI data sets featured sufficient spatial, spectral and temporal resolution to detect unique multi-temporal signatures for the region's major land cover types. It turned out that MODIS 250 NDVI time series data have high potential for large-basin land use/land cover monitoring and information updating for purposes of environmental basin research and management.

Rawat et al. (2013) illustrates an integrated approach of remote sensing and GIS (Geographical Information System), i.e. geospatial techniques for assessment of land use and land cover dynamics of a town located in the foothill zone of the Uttarakhand State viz., the Ramnagar. Landsat satellite imageries of two different time periods, i.e., Landsat Thematic Mapper (TM) of 1990 and 2010 were acquired by USGS Earth Explorer and quantified the land use and land cover changes in the Ramnagar town from 1990 to 2010 over a period of two decades. Supervised Classification methodology has been employed using Maximum Likelihood Technique in ERDAS 9.3. The images of the study area were categorized into five different classes, viz. built-up area, vegetation, agricultural land, water bodies and sand bar. The study reveals that the Ramnagar town is expanding maximum towards the southern direction along the National Highway-121. The study also highlights the importance of digital change detection techniques for nature and location of change of the Ramnagar Town area.

Vadrevu (2014) stated that two key challenges facing central Asian countries include land degradation and water resource management. Land cover change, unsustainable land use, and poor management of river waters in the region have created disputes. As the countries of Central Asia are heavily dependent on use of fragile dry lands and limited arable land, land-use change and water management are central issues in the region. A number of regional and international efforts have been made to understand the causes, extent, rate, and societal implications of land-use changes in the region, but these efforts have not been synthesized or framed effectively to address emerging issues. Common themes that surfaced during the discussions were the need to strengthen regional cooperation to solve trans boundary water-use issues in the region, and improve decision-making in the region informed through consensus scientific assessment. Participants identified the need to develop regionally consistent land-use and land-cover datasets.

Singh and Singh (2014) discussed that unprecedented growth of an area measured is an important task because it plays an important role for future development of that area. The change analysis was performed by post classification comparison method, comparing the data of two different sensors (Lands at TM and LISS III IRS P-6), at different time periods (1992 and 2004). The growth of Delhi measured between two time periods was based on the above data set. The results showed that there was rapid change in land cover/land use. It was found that there was a phenomenal change in the built-up area in watersheds, loss of forest cover and change in agriculture land. There is a great need for sustainable management of resources to maximize benefits of societal resources. On the basis of these data the land transformation map for different time periods were prepared and showing land transformation data.

Pandian (2014) stated that land use/ land cover is an important component in understanding the interactions of the human activities with the environment and thus it is necessary to monitor and detect the changes to maintain a sustainable environment. An attempt has been made to study the changes in land use and land cover parts of Coimbatore and Tiruppur districts. The study was carried out through

Remote Sensing and GIS approach using SOI toposheets, LANDSAT imagery of 2000 and IRS-P6-LISS-III 2009. The land use/land cover classification was performed based on the Survey of India toposheets and Satellite imageries. GIS software is used to prepare the thematic maps and ground truth observations were also performed to check the accuracy of the classification. The ten year time period of 2000-2009 shows the major type of land use change. The reasons for the change detection have been discussed in the study. However plantation, land with scrub, wet logged, barren rocky, tanks and reservoirs have also experienced the change. Coimbatore and Tiruppur district are identified as one of the industrialized areas in India. It is necessary to closely monitor the land use/land cover changes for maintaining a sustainable environment for a proper development.

The foregoing literature review reveals that majority of the studies have emphasized on the general nature of forest in different regions and only a few of them tries to focus on changes in forest cover and land use and land cover. The present study is a step in this direction.

1.4 STUDY AREA

The study area, Renuka forest division is situated in Sirmour district. It lies between 30°31′11″ and 30°52′16″ north latitudes and 77°17′34″ and 77°47′38″ east longitudes **(Fig. 1.1)**. It is bounded on the North by Chopal and Rajgarh Forest Divisions; on the East by Chakrata Forest Division of Uttarakhand; on the South by Nahan Forest Division and on the West by Paonta Sahib Forest Division (Working Plan of Forest, 2014). The geographical area of the division is 987 sq.km. Two wild life sanctuaries viz. Churdhar Wild Life Sanctuary and Renuka Wild Life Sanctuary also fall in the study area. There are five forest ranges in Renuka division namely Renuka, Sangrah, Nohra, Shillai and Kafota. The entire tract is mountainous and varies in elevation from 477 metre to 3647 metre above mean sea level. The slopes are generally steep to precipitous with deep stream and springs. The entire region of Renuka forest division falls within the catchments of Giri, Sainj and Tons rivers. The Jalal stream and Naitka rivulet are two important streams, which drains into Giri at Sieun and Khairi respectively (District Gazetteer, 1996).

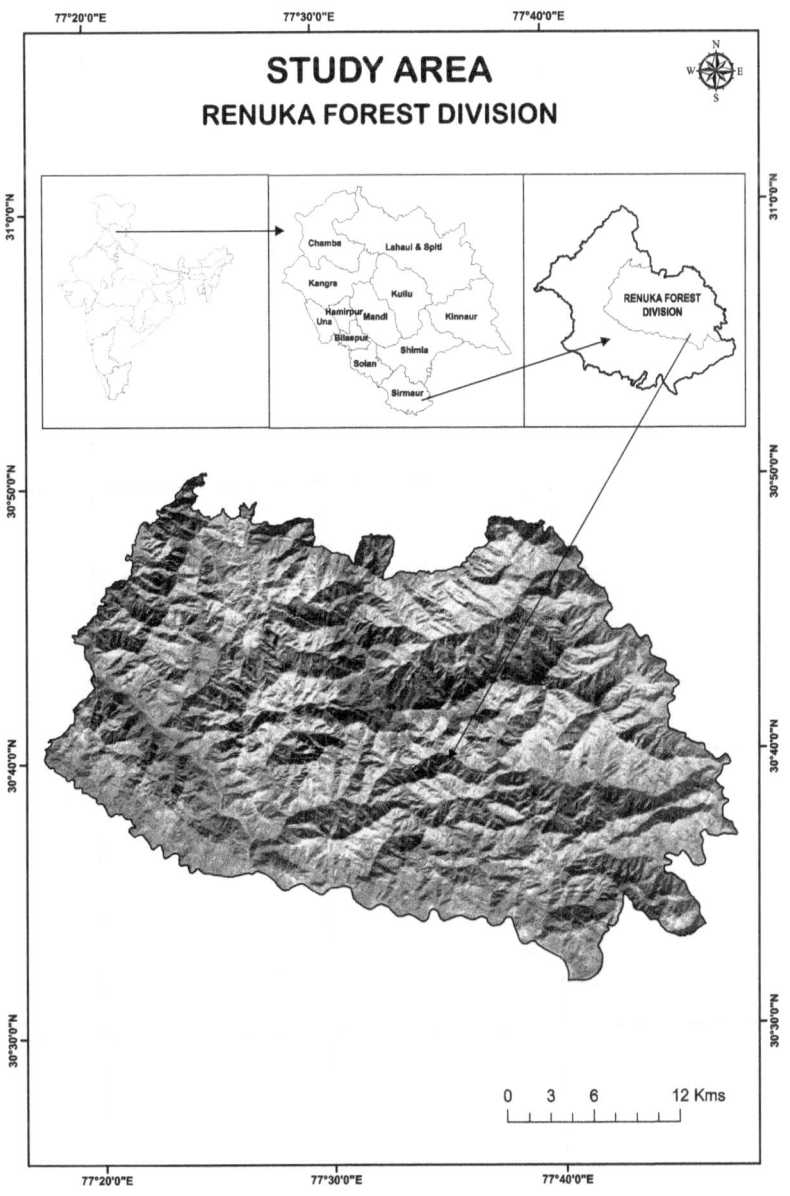

Fig. 1.1

Physiologically the tract forms the part of sub-Himalayan Zone and Lesser Himalayas. The sub-Himalayan zone extends to an altitude of 900 m. whereas the lesser Himalayan zone varies altitude from 900 m to 2900 m. The tract comprises of rocks ranging from Pre-Cambrian to Recent (Hayden, 1904). Soil is clayey loam, sandy clayey loam and clay in the study area. Temperature experiences both hot in summer as well as severe cold in winter and rainfall occurs maximum in summer and snow in winter in the higher altitude of the region. Renuka forest division has two tehsils; three sub tehsils and two community development blocks. The population of the study area is 1,67,705 (Census of India, 2011).

1.5 AIMS AND OBJECTIVES

The present study aims to analyze or to assess the impact of forest resources utilization on the natural environment. The basic objective is to analyze the forest cover and land use and land cover changes of the study region so that an appropriate planning for sustainable development can be evolved. In order to achieve this objective understanding of natural environment, its health and vigour is essential. Considering all these factors the present study aims to achieve the following objectives.

1. To assess the spatial and temporal variation of forest cover and changes.

2. To study the modes adopted to extract the forest resources and their consequences.

3. To analyse the spatial and temporal variation of land use and land cover changes.

4. To understand the people's perception of land use / land cover and forest sector policies.

1.6 MAJOR HYPOTHESES

1. The forest cover has changed drastically in the study area.

2. The forest depletion has greatly influenced the physical environment of the study area.

3. The forests are degrading due to growing population and their multifarious activities.

1.7 DATA SOURCES

Present study is primarily based on a harmonious production of both primary and secondary data. Both the quantitatively as well as qualitative techniques have been used to analyze the data. Survey of India topographical sheets, State of Forest Reports of Forest Survey of India and satellite imageries of the study area has been used extensively along with other collateral data which has been obtained from various professional journals, doctoral theses and reports of various government departments viz. Forest Deptt., Economics and Statistical Deptt., Census Deptt., some governmental and non-governmental agencies etc. Extensive field work has been done by conducting structured questionnaires, schedules, interviews and personnel observations in order to collect primary data.

Following are the major sources of data used in the present study:

1. Survey of India topographical sheets no. 53F/5, 53F/6, 53F/9 and 53F/10 on 1:50,000 scale.

2. State of Forest Reports of Forest Survey of India.

3. Satellite imageries of Landsat and Resource Sat-II of different time period.

4. Census data obtained from Directorate of Census, Himachal Pradesh.

5. Data gathered from various published and unpublished official reports.

6. Data gathered through questionnaires, schedules interviews and personal observations.

1.8 METHODOLOGY

To establish the relationship between human and forest resources and their ecology in the area under study require intensive and extensive fieldwork. Thus, lot of data has been generated through rigorous field surveys, by preparing questionnaires/schedules and by interviewing the people so that it becomes possible to establish a cause and effect relationship. The household level data has been collected from 100 villages out of 156 villages which are well spread over the forest division. More than half of the villages have been visited and data has been collected through structured questionnaire and schedule. This has been done by random selection of the villages and households after stratifying the study area on the basis of forest cover i.e.

dense, open, bushes and no forests area. To understand the forest ecology and principles of its management, the present study envisages undertaking the study of interaction between forests on the one hand and human population on the other.

The study was mainly based on collection and analysis of remote sensing data. Ground surveys were also used to collect data on forest cover and land use/cover and to verify the results of visual interpretation of satellite imagery. A time-series of Landsat MSS, Thematic Mapper (TM), and Enhanced Thematic Mapper Plus (ETM+) images were used to derive forest cover and land use/ cover maps of the area. The dataset included full scenes for the years 1972, 1989, 2001, and 2011. The selected datasets were cloud free images acquired between March and May, i.e. spring-time when vegetation growth or cover is at its peak. The dataset was mainly downloaded from the archive of the Global Land Cover Facility (GLCF) (http://glcf. umiacs.umd.edu/index.shtml). Resource Satellite-II imagery of 2011 from NRSC, Hyderabad has also been used.

Remote sensing and GIS techniques have been employed in order to carry out a scrutinized study and to procure more accurate and real time information. With the help of these geospatial tools baseline studies has been conducted to acquire information regarding physio-cultural condition of the study area. Various thematic maps has been prepared using appropriate cartographic methods. Data gathered through primary sources was computed carefully and displayed with the help of tables, graphs, diagrams and maps.

Monitoring the forest status and assessing impacts caused by human activities are important for planning purpose and effectiveness of forest management. With the advent of the spatial data acquisition using GIS and remote sensing technology, the assessment of forest resources are now becoming more sophisticated and accurate. These technologies have enabled generating knowledge of biophysical and socio economic aspects of forest resources. They have been extensively used in mapping resources; land cover and land uses. Furthermore, they allow us to develop spatial models that are useful in translating impacts prediction results into appropriate management plan and policy measures.

1.9 ORGANIZATIONAL FRAMEWORK

The present thesis has been prepared under seven chapters.

The **first chapter** is introductory which gives statement of the problem, review of literature, study area, objectives, hypotheses, data sources and methodology.

The **second chapter** present the theoretical background of the study and discuss concept of natural resources, their classification, concept of forest resource, forest products, forest conservation, National forest policies and State forest policy etc.

The **third chapter** dwells upon the physio-cultural setting of the study area. It highlights the physical and socio-economic feature of the study area.

The **fourth chapter** has been devoted to spatial distribution of forest resources in Renuka forest division. It analysed the forest types, classification of vegetation, early history of the forests, forest cover and changes of the study area and also give people's perception about forest cover changes.

The human impact on physical environment has been analyzed in the **fifth chapter**. This chapter focuses on the mpact of man on the physical environment, forest fire, unscientific mining or quarrying, exploitation of non-timber product, medicinal plants, resin tapping, livestock grazing and people's perception about forest sector policies.

The **sixth chapter** studies the land use/ land cover pattern in Renuka forest division; concept of land use/ land cover pattern and classification criteria, land use/ land cover pattern of Renuka division 1972, land use/ land cover pattern of Renuka division 1989, land use/ land cover pattern of Renuka division 2001, land use/ land cover pattern of Renuka division 2011, land use and land cover changes in Renuka forest division and people's perception about land use/ land cover pattern.

The summary, conclusions and policy imperatives forms the part of the **seventh and final chapter**.

REFERENCES

Achard, F. et al. (2002): Determination of Deforestation Rates of the World's Humid Tropical Forests, *Science Magazine*, 297(55): PP 999-1002.

Adeniyi, P.O. and Omojola, A. (1999): Land Use Land Cover Change Evaluation in Sokoto-Rima Basin of North Western Nigeria Based on Archival of the Environment (AARSE) on Geoinformation Technology Applications for Resource and Environmental Management in Africa, PP 143-172.

Anderson, et al. (1976): A Land Use and Land Cover Classification System for Use with Remote Sensor Data, *Geological Survey Professional*, U.S. Government Printing Office, Washington, D.C. PP 28.

Anon, (1988): National Forest Policy 1988, Ministry of Environment and Forests, Government of India.

Annual Report (2012-13): Department of Animal Husbandry, *Dairying and Fisheries, Ministry of Agriculture*, Govt. of India.

Aswal, B.S. and Mehrotra, B.N. (1994): Flora of Lahaul-Spiti: A Cold Desert in North-West Himalayas. Bishen Singh, Mahender Pal Singh, Dehra Dun and Periodical Express, Delhi, PP 761.

Awasthi, K.D., Sitaula, B.K., Singh, B.R. and Bajracharya, R.M. (2002): Land Use Change in Two Nepal's Watershed: GIS and Geomorphometric Analysis, Land Degradation and Development, 13: PP 495-513.

Awasthi, K.D. (2004): Land Use Change Effects on Soil Degradation, Carbon and Nutrient Stocks and Greenhouse Gas Emission in Mountain Watershed, Ph.d Thesis, Agriculture University, Norway.

Awasthi, K.D. et al. (2005): Analysis of Land Use Structure in Two Mountain Watersheds of Nepal Using FRAGSTATS, *Forestry: A Journal of Forestry Nepal,* 13(6): PP 495-513.

Bazzaz, F.A. (1975): Plant Species Diversity in Old Field Successional Ecosystems in Southern Illinois, *Ecology,* 56: PP 485-488.

Bhattrai, K. and Dennis, C. (2007): Evaluating Land Use Dynamics and Forest Cover Change in Nepal's Bara District (1973-2003), *Journal Human Ecology,* PP 81-95.

Bharti, R.R., Ishwar, D.R., Adhikar, B.S. and Rawat, G.S. (2011): Timberline Change Detection Using Topographic Map and Satellite Imagery: A Critique, *Tropical Ecology,* 52(1): PP 133-137.

Birendra, K.C. and Shin, N. (2006): Refugee Impact on Collective Management of Forest Resources: A Case Study of Bhutanese Refugees in Nepal's Eastern Terai Region, *Journal of Forest Research,* 11(5): PP 305-311.

Bliss, L.C. (1966): Plant Productivity in Alpine Micro-Environment on Mountain Washington, New Hampshire, *Ecology Monograph,* 36: PP 125-135.

Brent, C.J. and Peter, R.B. (2014): Drivers of Change in Landholder Capacity to Manage Natural Resources, *Journal of Natural Resources Policy Research,* 6(1): PP 1-26.

Census of India (2011): Census of India, *Ministry of Home Affairs,* Govt. of India.

Champion, H.G. and Seth, S.K. (1968): A Revised Survey of the Forest Types of India, *The Manager of Publication,* New Delhi.

Chaturvedi, A.N. (1981): Big Forest Loss Every Minute, *The Daily National Heraled,* Lucknow, October 19, 1981.

Chaturvedi, O.P. and Singh, J.S. (1983): Estimation of Biomass and Biomass Production of Pinus Roxburghii Tree Growing in All Aged Natural Forests, *Canadian Journal of Forest Research,* 12(3): PP 632-640.

Cheng, S. et al. (1998): Deforestation and Degradation of Natural Resources in Ethiopia: Forest Management Implications From A Case Study in the Belete-Gera Forest, *Journal of Forest Research,* 3(4): PP 199-204.

Chun, J. (2014): A Legal Approach to Induce the Traditional Knowledge of Forest Resources, *Forest Policy and Economics,* 38: PP 40-45.

Chytry, M. (2001): Phytosociological Data Give Biased Estimates of Species Richness, *Journal of Vegetation Science,* 12(3): PP 439-444.

Coppin, P.P. and Bauer, M.E. (1994): Processing of Multitemporal Landsat TM Imagery to Optimize Extraction of Forest Cover Change Features. *IEEE Transactions of Geoscience and Remote Sensing,* 32(4): PP 918-927.

Coppin, P. and Bauer, M. (1996): Digital Change Detection in Forest Ecosystems with Remote Sensing Imagery, *Remote Sensing Reviews,* 13: PP 207-234.

Clark, A.N. (2003): Dictionay of Geography, Penguin Books.

Daniel, et al. (2002): A Comparison of Land Use and Land Cover Change Detection Methods, ASPRS-ACSM Annual Conference and FIG XXII Congress, PP 2.

Dansereau, P. (1960): Origin and Growth of Plant Communities, In: Growth in Living Systems Proceedings of International Symposium Held at Purdue University, Basic Book Inc., New York, PP 567-603.

Desclee, B. (2007): Automated Object-Based Change Detection for Forest Monitoring by Satellite Remote Sensing: Applications in Temperate and Tropical Regions, Ph.D Thesis.

Desclee, B., Bogaert, P. and Defourny, P. (2006): Forest Change Detection by Statistical Object-Based Method, *Remote Sensing of Environment,* 102: PP 1-11.

Dongsheng, L. et al. (2004): Forest Resources and Environment in China, *Journal of Forest Research,* 9(4): PP 307-312.

Dimyati (1995): An Analysis of Land Use and Land Cover Change Using the Combination of MSS Landsat and Land Use Map-A Case Study of Yogyakarta, Indonesia, *International Journal of Remote Sensing,* 17(5): PP 931-944.

District Gazetteer (1996): Gazetteer of the Sirmour District, Indus Publishing Company.

Dwivedi, A.P. (1993): Forests, the Ecological Ramifications, Natraj Publishers, Dehradun, PP 14-15.

EOSAT (1992): Landsat TM Classification International Georgia Wetlands in EOSAT Data User Notes, EOSAT Company, Lanham, MD, (7)1.

EOSAT (1994): EOSAT's Statewide Purchase Plan Keeps South Carolina Residents in the Know, in EOSAT Notes, EOSAT Company Lanham, MD, 9(1).

Ellis, E. and Pontius, Jr. (2006): Land-Use and Land Cover Change Encyclopedia of Earth. Available At: Http://Www.Eoearth.Org/Article/Land-Use_And_Land-Cover_Change.

Ellum, D.S. (2009): Floristic Diversity in Managed Forests: Demography and Physiology of Understory Plants Following Disturbance in Southern New England Forests, *Journal of Sustainable Forestry,* 28: PP 132-151.

ERDAS Field Guide (1999): Earth Resources Data Analysis System, *ERDAS Inc. Atlanta,* Georgia, PP 628.

FAO (1992): The Forest Resources of the Tropical Zone by Main Ecological Regions, Forest Resources Assessment Project, FAO Rome, PP 216-411.

FAO (2000): The Global Forest Resources Assessment 2000 (FRA 2000), Food and Agriculture Organization of the United Nations, Rome, Italy.

Fernandez, N.R., Rodriquez, T.C., Areguin Sanchez, M., Dela, A. and Rodriguez, J.A. (1996): Floristic List of the Basal River Basin, Mexico, Poliibortanica, No. 9: PP 151.

Fitzpatrick, L. et al. (1987): Producing Alaska Interim Land Cover Maps from Landsat Digital and Ancillary Data, In Proceedings of the 11th Annual William T. Pecora Memorial Symposium: Satellite Land Remote Sensing: Current Programs and A Look to the Future, *American Society of Photogrammetry and Remote Sensing,* PP 339-347.

Forest Statistics (1998): Forest Department, Himachal Pradesh.

Franklin, J., Phinn, S.R., Woodcock, C.E., and Rogan, J. (2003): Rationale and Conceptual Framework for Classification Approaches to Assess Forest Resources and Properties, In: Wulder, M. A. and Franklin, S. E (eds) Remote Sensing of Forest Environments Concepts and Case Studies, Kluwer Academic Publishers, USA.

Franklin, S. E., Moskal, M.B. and La Ugh. (2000): Interpretation and Classification of Partially Harvested Forest Stands in the Fundy Model Forest Using Multitemporal Landsat TM Digital Data, *Canadian Journal of Remote Sensing,* 26(4): PP 318-333.

Franklin, S.E., Lavigne, M.B., Wulder, M. A., and Mccaffrey, T.M. (2002): Large Area Forest Structure Change Detection: An Example, *Canadian Journal of Remote Sensing,* 28(4): PP 588-592.

Franklin, S.E., Lavigne, M.B., Wulder, M.A. and Stenhouse, G.B. (2002): Change Detection and Landscape Structure Mapping Using Remote Sensing, *The Forestry Chronicle,* 78(5): PP 618-625.

Franklin, S., Lavigne, M., Wulder, M. and Mccaffrey, T. (2002): Large Area Forest Structure Change Detection: An Example, *Canadian Journal of Remote Sensing,* 28(4): PP 588-592.

FRI (1995): Forestry Statistics of India (1995): Indian Council for Forestry Research and Education, Dehradun.

FSI (1996): Fuel Wood, Timber and Fodder from Forests of India, Government of India, *Ministry of Environment and Forest,* Dehradun.

FSI (1999): The State of Forest Report 1999, Government of India, Forest Survey of India, *Ministry of Environment and Forest,* Dehradun.

Fuller, D.O. (2001): Forest Fragmentation in Loudoun County, Virginia, USA Evaluated with Multi-temporal Landsat Imagery, *Landscape Ecology,* 16: PP 627-642.

Gadgil, M. (1994): Inventoring Monitoring and Conserving India's Biological Diversity, *Current Science,* 66(6): PP 401-406.

Gadgil, M. and Guha, R. (1995): Ecology and Equity: The Use and Abuse of Nature in Contemporary India, Penguin Books India.

Gaston, A.J., Hunter, M.L. and Garson, P.J. (1981): The Wildlife of Himachal Pradesh, Western Himalayas, University of Maine, *Technical Report* No. 82, Maine, USA.

Gentry, A.H. (1988): Changes in Plant Community Diversity and Floristic Composition on Environmental and Geographical Gradients, *Annals of the Missouri Botanical Garden,* 75: PP 1-34.

Gupta, A.C., Raturi, D.P. and Bora, N.K.S. (1982): Floristic Composition of a West Himalayan Subalpine Coniferous Forests, *Van-Vigyan*, 20(2): PP 51-54.

Gupta, R., Chaudhary, P.R. and Wate, S.R (2008): Ethnomedicinal Uses of Some Plants Used by Gond Tribe of Bhandara District, Maharashtra, *Indian Journal of Traditional Knowledge*, 9(4): PP 713-717.

Hansen, et al. (2013): High-Resolution Global Maps of 21st-Century Forest Cover Change, *Science Magazine*, 342 (6160): PP 850-853.

Hansen, M.J., Franklin, S.E., Woudsma, C. and Peterson, M. (2001): Forest Structure Classification in the North Columbia Mountains Using the Landsat TM Tasseled Cap Wetness Component, *Canadian Journal of Remote Sensing*, 27(1): PP 20-32.

Hayden, H.H. (1904): The Geology of Spiti with Parts of Bashar and Rupshu. Mem. Geological Survey India, 36(1).

Heywood, V.H. (1995): Global Biodiversity Assessment, Cambridge University Press.

Hurlbert, S.H. (1971): The Non-Concept of Species Diversity: A Critique and Alternative Parameters, *Ecology*, 52: PP 577-586.

Hwan, K.C., Hyum, M. and Cheul, S.B. (1999): Species Diversity of Forest Vegetation in Mt. Jangan, Ehollobuk-Do-Korean, *Journal of Environment and Ecology*, 13(3): PP 271-279.

International Union of Conservation of Nature and Natural Resources (IUCN) (1994): IUCN Red List Categories, IUCN, Gland. Inventory, *Journal of Forestry*, 97: PP 44-48.

Jain, S.K. and Srivastava, A. (1994): On the Status of Endemism of Some More Grasses in India, *Indian Journal of Forestry*, 17(1): PP 89-90.

Joshi, P.K. et al. (2001): Forest Cover Assessment in Western Himalayas, Himachal Pradesh Using IRS 1C/1D Wifs Data, *Current Science*, 8(80): PP 941-947.

Kennedy, R.E., Cohen, W.B. and Schroeder, T.A. (2007): Trajectory Based Change Detection for Automated Characterization of Forest Disturbance Dynamics, *Remote Sensing of Environment*, 110: PP 370-386.

Khullar, D.R. (2010): India: A Comprehensive Geography, Kalyani Publishers, New Dehli.

Kumar, D. et al. (2012): Ethno-Medicinal Uses of Some Plants of Kanag Hill in Shimla, Himachal Pradesh, India, *International Journal of Research in Ayurveda and Pharmacy*, 3(2): PP 319-322.

Kumar, J.S., Arockiasamy, D.I. and Britto, J.S. (2002): Forest Type Mapping and Vegetation Analysis in Part of Kolli Hills, Eastern Ghats of Tamil Nadu, *Tropical Ecology,* 43: PP 345-349.

Kumar, N. and Choyal, R. (2013): Ethno-Medicinal Uses of Some Plants of Lower Foot Hills of Himachal Pradesh for the Treatment of Oral Health Problems and Other Mouth Disorders, *International Journal of Advanced Research,* 1(5): PP 1-7.

Kushwaha, S.P.S. (1985): Environmental Monitoring and Cyclone Impact Assessment on Sriharikota Island, India, Project Report, *National Remote Sensing Agency,* Hyderabad.

Kushwaha, S.P.S. (1990): Forest Type Mapping and Change Detection from Satellite Imagery, *ISPRS Journal of Photogrammetry and Remote Sensing,* 45: PP 175-181.

Kushwaha, S.P.S., Kuntz, S. and Oesten, G. (1994): Applications of Images Texture in Forest Classification, *International Journal of Remote Sensing,* 15(11): PP 2273- 2284.

Lambin, E.F. et al. (1999): Land Use and Land Cover Change (LULCC): Implementation Strategy, IGBP Report No. 48, IHDP Report No. 10, Stockholms, Bonn.

Lambin, E.F. et al. (2001): The Causes of Land Use and Land Cover Change: Moving Beyond the Myth, *Global Environment Change,* 11: PP 261-269.

Lal, J.B. (1989): India's Forests: Myth and Reality, Natraj Publication Dehra Dun, PP 17-26.

Lugo, A.E. (1978): Stress And Ecosystem, In: Energy And Environmental Stress in Aquatic Systems, J.H. Thorp and J.W. Gibbons (Eds.), DOE Symposium Series, *National Technical Information Service,* Springfield, PP 62-101.

Macleod, R.D. and Congalton, R.G. (1998): A Quantitative Comparison of Change Detection Algorithms for Monitoring Eelgrass from Remotely Sensed Data, *Photogrammetric Engineering and Remote Sensing,* 64(3): PP 207-216.

Masek, J.G., Huang, C., Cohen, W.B., Kutler, J., Hall, F.G. and Wolfe, R. (2008): North American Forest Disturbance Mapped from a Decadal Landsat Record: Methodology and Initial Results, *Remote Sensing of Environment,* 112: PP 2914-2926.

Mahajan, K.K. (1983): Conservation of Biological Diversity in the Himalayas, J.S. Singh 1985, *Central Himalayan Environment Association,* PP 440-447.

Margalef, R. (1968): Perspectives in Ecological Theory, University Chicago Press, Chicago, PP 78-87.

Mas, J.F. (1999): Monitoring Land Cover Changes: A Comparison of Change Detection Techniques, *International Journal of Remote Sensing,* 20(1): PP 139-152.

Mertens, B., Sunderlin, W.D. and Lambin, E.F. (2000): Impacts of Macro- Economic Change on Deforestation in South Cameroon: Integration of Household Survey and Remotely Sensed Data, *World Development,* 28(6).

Meyer, W.B. (1995): Past and Present Land Use and Land Cover in the USA Consequences, PP 24-33.

Meyer, W.B. and Turner II, L.B. (1992): Human Population Growth and Global Land Use and Land Cover Change, *Journal Annual Review of Ecology and Systematics,* 23: PP 29-61.

Mitchell, B. (1989): Geography and Resource Analysis, Longman, London and New York.

Miah, M.D. et al. (2012): Contribution of Forests to the Livelihood of the Chakma Community in the Chittagong Hill Tracts of Bangladesh, *Journal of Forest Research,* 17(6): PP 449-457.

Mishra, R. (1968): Community Structure, In: Ecological Work Book, Oxford and IBH Publishing Company, Calcutta, 16: PP 31-68.

Mitchard, E.T.A., Saatchi, S.S., Gerard, S.L.L. and Meir, P. (2009): Measuring Woody Encroachment from 1982-2006 along a Forest-Savanna Boundary in Central Africa, *Earth Interactions,* 18(8).

Moshen, A. (1999): Environmental Land Use Change Detection and Assessment Using with Multi-Temporal Satellite Imagery, Zanjan University.

Nautiyal, S. (1991): Willow a Multipurpose Tree of Cold Desert of India, *Indian Forester,* 117(2): PP 153-155.

Nautiyal, S., Pandita, V.K. and Naithani, H.B. (1994): Exploration and Collection of Fodder Grasses and Legumes Germplasm from Cold Desert of India, *Annals of Forestry,* 2(1): PP 77-79.

Newsweek (1982): Where All the Forests Gone, Newsweek, October 6, 1982.

Nicholson, S.A. and Monk, C.D. (1974): Plant Species Diversity in Old-Field Succession on the Georgia Piedmont, *Ecology,* 5: PP 1075-1085.

NRSA (1983a): Report on Environmental Studies through Remote Sensing and Mapping of Idukki Area in Kerala, *National Remote Sensing Agency,* Hyderabad.

NRSA (1983b): Nationwide Mapping of Forest and Non-Forest Areas Using Landsat False Colour Composites for the Periods, 1972-75 and 1980-82, Report, National Remote Sensing Agency, Hyderabad, on Satellite Remote Sensing, *Journal of Indian Society of Remote Sensing*, 18: PP 29-42.

Olorunfemi, J.F. (1983): Monitoring Urban Land Use in Developed Countries- An Aerial Photographic Approach, *Environmental International*, 9: PP 27-32.

Opeyem, A.Z. (2006): Change Detection in Land Use and Land Cover Using Remote Sensing Data and GIS: A Case Study of Ilorin and its Environs in Kwara State, Msc Theisis, Department of Geography of Ibadan, Ibadan.

Ota, T. et al. (2011): Influence of Using Texture Information in Remote Sensed Data on the Accuracy of Forest Type Classification at Different Levels of Spatial Resolution, *Journal of Forest Research,* 16(6): PP 432-437.

Pandey, A.C. and Nathawat, M.S. (2006): Land Use and Land Cover Mapping through Digital Image Processing of Satellite Data- A Case Study from Panchkula, Ambala and Yamunanagar Districts, Haryana State, India.

Pandian, M. et al. (2014): Land Use and Land Cover Change Detection Using Remote Sensing and GIS in Parts of Coimbatore and Tiruppur Districts, Tamil Nadu, India, *International Journal of Remote Sensing and Geoscience,* 3(1): PP 15-20.

Panigrahy, S.D. et al. (2010): Timberline Change Detection Using Topographic Map and Satellite Imagery, *Tropical Ecology,* 51: PP 87-91.

Phong, L.T. (2004): Analysis of Forest Cover Dynamics and Their Driving Forces in Bach Ma National Park and Buffer Zone Using Remote Sensing and GIS, Msc Thesis, ITC, Enschede, the Netherland.

Phung, T.T. et al. (2014): Impacts of Changes in Mangrove Forest Management Practices on Forest Accessibility and Livelihood: A Case Study in Mangrove-Shrimp Farming System in Ca Mau Province, Mekong Delta, Vietnam, *Science Direct,* 36: PP 89-101.

Pielou, E.C. (1975): Ecological Diversity, New York.

Pradhan, B. and Awang, M.A.B. (2007): Forest Fire Susceptibility and Risk Mapping Using Remote Sensing and Geographical Information Systems (GIS), *Disaster Prevention and Management,* 16(3): PP 344-352.

Prakasam, C. (2010): Land Use and Land Cover Change Detection through Remote Sensing Approach: A Case Study of Kodaikanal Taluk, Tamil Nadu, *International Journal of Geomatics and Geosciences,* 1(2).

Ralhan, P.K., Saxena, A.K. and Singh, J.S. (1984): Analysis of Forest Vegetation at and Around Nainital in Kumaun Himalaya, *Indian National Science Academy,* 49(1): PP 121-137.

Ram, P. Pandey, R.K. and Prasad, R. (1992): An Observation on Plant Diversity of Sal and Teak Forest in Relation to Intensity of Biotic Impact of Various Distances from Habitation in Madhya Pradesh- A Case Study, *Journal of Tropical Forester,* 8(1): PP 62-83.

Ramande, F. (1984): Ecology of Narural Resources, Jonn Wiley and Sons, New York.

Rawat, et al. (2013): Changes in Land Use and Land Cover Using Geospatial Techniques: A Case Study of Ramnagar Town Area, District Nainital, Uttarakhand, India, *The Egyptian Journal of Remote Sensing and Space Sciences,* 16: PP 111-117.

Read, J.M. and Lam, N.S.N. (2002): Spatial Methods for Characterizing Land Cover and Detecting Land Cover Changes for the Tropics, *International Journal of Remote Sensing,* 23(12): PP 2457-2474.

Riebsame, W.E., Meyer, W.B. and Turner II, B.L. (1994): Modeling Land Use and Land Cover as Part of Global Environmental Change, *Climate Change,* 28: PP 45-64.

Roy, P.S. (1993): Remote Sensing for Forest Ecosystem Analysis and Management, Environmental Studies in India, ed. M. Balakrishnan, Published by Oxford and IBH, New Delhi, PP 335-363.

Roy, P.S., Dutt, C.B.S. and Joshi, P.K. (2000): Tropical Forest Resource Assessment and Monitoring, *Tropical Ecology,* 43(1): PP 21-37.

Sader, A., Bertrand, M. and Wilson, E. (2003): Satellite Change Detection of Forest Harvest Patterns on an Industrial Forest Landscape, *Forest Science,* 49(3): PP 341-353.

Sader, S. A., Hayes, D.J., Hepinstall, J.A., Coan, M.C., and Soza, C. (2001): Forest Change Monitoring of a Remote Biosphere Reserve, *International Journal of Remote Sensing,* 22(10): PP 1937-1950.

Sader, S. and Winne, J. (1992): RGB-NDVI Color Composites for Visualizing Forest Change Dynamics, *International Journal of Remote Sensing,* 13(16): PP 3055-3067.

Sala, O.E. et al. (2000): Biodiversity: Global Biodiversity Scenarios for the Year 2100, *Science,* 287: PP 1770-1774.

Sakthivel. R, Manivel, J.R., Pugalanthi, R. and Vijay, A. (2010): Remote Sensing and GIS Based Forest Cover Change Detection Study in Kalrayan Hills, Tamil Nadu, *Journal of Environmental Biology,* 31(5): PP 737-747.

Shafi, M. and Raza, M. (1992): Forest Ecosystems of the World, Rawat Publications, New Delhi.

Sharma, A. et al. (2010): A Base Line Survey of Protected Area Network in North Bank Landscapes (NBL) with Reference to Land Use Land Cover Change (LULCC) Using Remote Sensing, 13th Annual International Conference and Exhibition on Geospatial Information Technology and Application, WWF-India.

Schnitzler, A. (1994): Conservation of Biodiversity in Alluvial Hardwood Forests of the Temperate Zone, *Forest Ecology and Management*, 68: PP 385-398.

Schweinfurth, U. (1968): Vegetation of the Himalayan Mountain and Rivers of India *21st International Georaphy Congress,* New Delhi, PP 110-136.

Shackleton, C.M. and Pandey, A.K. (2014): Positioning Non-Timber Forest Products on the Development Agenda, *Forest Policy and Economics,* 38: PP 1-7.

Shafi, M.I. and Yarronton, G.G.A. (1973): Diversity Floristic Richness and Species Evenness during a Secondary (Post-Fire) Succession, *Ecology,* 54: PP 897-902.

Sharafeldin, M.A. (1982): Nomadic Systems in I. E. Coop (Ed), Sheep and Goat Production, World Animal Sciences, CI. Elsevier Scientific Publishing Co. Amesterdam, Neatherlands.

Shosheng and Kutiel (1994): Monitoring Temporal Vegetation Cover Changes in Mediterranean and Arid Ecosystems Using a Remote Sensing Technique: Case Study of the Judean Mountain and the Judean Desert, *Journal of Arid Environments,* 10(6).

Singh, A. (1989): Digital Change Detection Techniques Using Remotely Sensed Data, *International Journal of Remote Sensing,* 10(6): PP 989-1003.

Singh, B. and Singh, J. (2014): Land Use and Land Cover Change of Delhi: A Study Using Remote Sensing and GIS Techniques, *International Research Journal of Earth Sciences*, 2(1): PP 15-20.

Singh, J.S. and Singh, S.P. (1987): Forest Vegetation of Himalaya, *Botanical Review,* 53(1): PP 80-192.

Singh, J.S. and Singh, S.P. (1992): Forests of the Himalayas, Ganodaya Prakashan, Nanital, India.

Singh, N.B. (1990): Medicinal Wealth of Kinnaur, *Van Vigyan,* 28(4): PP 174-181.

Singhal, R.M. et al. (2003): Forests and Forestry Research in India, *Tropical Ecology,* 44(1): PP 55-61.

Singh, S.P. (1998): Assessment of Floral and Habit Diversity and Collection of Baseline Data to Monitor Vegetation of GHNP and Wls. WII, Report Dehradun.

Singh, S.K. (1998): Vegetation Structure under North and South Aspects in the Temperate Zone of Tirthan Valley, Western Himalayas, *Indian Journal of Forestry,* 21(3): PP 217-223.

Skole, D. and Tucker, C. (1993): Tropical Deforestation and Habitat Fragmentation in the Amazon: Satellite Data from 1978 to 1988, *Science,* 260: PP 1905-1909.

Smith, N. (1981): Wood: an Ancient Fuel with a New Future, World Watch Papers 42, World Watch Institute, Washington, U.S.A.

Solangaarachchi, S.M. and Perera, B.M.S. (1993): Floristic Composition and Medicinally Important Plants in the Understorey of the Tropical Dry Mixed Evergreen Forest at the Hurulu Reserve of Sri Lanka, *Journal of the National Science Council of Sri Lanka,* 21(2): PP 209-226.

Song, K. et al. (2011): Land Use and Land Cover (LULC) Classification with MODIS Time Series Data and Validation in the Amur River Basin, *Geography and Natural Resource,* 32(1): PP 9-15.

State of Forest Report (SFR) (2001): Forest Survey of India, Ministry of Forest and Environment, Dehradun.

State Forest Report (SFR) (2011): Forest Survey of India, Ministry of Forest and Environment, PP 139-142.

Stehman, S.V. et al. (2003): Thematic Accuracy of the 1992 National Land Cover Data for the Eastern United States: Statistical Methodology and Regional Results, *Remote Sensing of Environment,* 86: PP 500-516.

Stibig, H.J. et al. (2014): Change in Tropical Forest Cover of Southeast Asia from 1990 to 2010, *Biogeosciences,* 11: PP 247-258.

Tansley, A.G. (1920): The Classification of Vegetation and the Concept of Development, *Journal Ecology,* 8: PP 118-149.

Tansley, A.G. (1947): British Ecology during the Past Quarter Century: the Plant Community and the Ecosystem, *Journal Ecology,* 27: PP 513-534.

Tiwari, M.K. and Saxena, A. (2011): Change Detection of Land Use/ Land Cover Pattern in an Around Mandideep and Obedullaganj Area, Using Remote Sensing and GIS, *International Journal of Technology and Engineering System,* 2(3).

Townshend, J.R.G. et al. (1995): The NASA Landsat Pathfinder Humid Tropical Deforestation Project, Land Satellite Information in the Next Decade (Tyson's Corner, Virginia: American Society of Photogrammetry and Remote Sensing), PP 76-87.

Turner II, et al. (1990): The Earth as Transformed by Human Action: Global and Regional Changes in the Biosphere over the Past 300 Years, Cambridge University Press, Cambridge.

Turner II, et al. (1995): Land Use Land Cover Change Science/ Research Plan, Joint Publication of the International Geosphere- Biosphere Programme (Report No. 35) and the Human Dimension of the Global Environmental Change Programme (Report No. 7), Royal Swedish Academy of Sciences, Stockholm, Sweden.

UNEP (2001): State of the Environment 2001, United Nation Environment Programme, Regional Resource Center for Asia and the Pacific (UNEP-RRC.AP), Thailand, UNFAO 1993, Forest Resource Assessment 1990, Tropical Countries Forestry Paper 112, Food and Agriculture Organization, Rome.

Vadrevu, K.P. (2014): NASA LCLUC Science Team Meeting on Land Use and Water Resources in Central Asia, *The Earth Observer*, 26(2): PP 14-18.

Vitousek, P.M. (1994): Beyond Global Warming: Ecology and Global Change, *Ecology*, 75: PP 1861-1876.

Walter, H. (1979): The Vegetation of the Earth and the Ecological Systems of the Geo-Biosphere, *Springer Verlag*, New York, PP 274.

Whittaker, R.H. (1969): Evolution of Diversity in Plant Communities, *Brookhaven Symposia in Biology*, 22: PP 178-196.

Whittaker, R.H. (1975): Communities and Ecosystems, 2nd edition, Macmillan Publication Company, New York, PP 385.

Whittaker, R.H. (1975): Communities and Ecosystems, Mcmillan, New York.

Wilson, E.O. (1994): The Diversity of Life, Penguin Books, Ltd.

Woodcock, C., Macomber, S., Pax-Lenney, M., and Cohen, W. (2001): Monitoring Large Areas for Forest Change Using Landsat: Generalization Across Space, Time and Landsat Sensors, *Remote Sensing of Environment*, 78(1-2): PP 194-203.

Working Plan of Forest (2014): Working Plan for the Forests of Renuka Forest Division, H.P. Govt., Forest Department.

World Bank (1980): Firewood Crops: Shrubs and Tree Species for Energy Production, National Academy of Sciences, Washington DC, U.S.A. PP 12-15.

World Conservation Monitoring Centre (WCMC) (1992): Global Diversity: Status of Earths Living Resources, World Conservation Monitoring Centre, Cambridge, Chapman and Holl, London, U.K.

World Resource Institute (WRI) (1989): Keeping Options Alive: The Scientific Basis for Conserving Biodiversity, In: Biological Diversity and Developing Countries- Issues and Options, A Synthesis Paper (Michael Flint), ODA, PP 50.

Wuenscher, J.E. (1974): The Ecological Niche and Vegetation Dynamics, In: Hand Book of Vegetation Science, Part-VI, Vegetation and Environment (ed), Br. Starin and W.D.W.D. Billings, Junk, the Hague, PP 98-145.

Wynne, R.H. and Carter, D.B. (1997): Will Remote Sensing Live up to its Promise for Forest Management, *Journal of Forestry,* 95: PP 23-25.

Yadav, K. (2010): Timber Forest Certification Scenario: An Overview, in K. Jayaraman, M. Balasundaran, K.V. Bhat and C.N. Krishnankutty (eds) Production And Marketing of Teakwood: Future Scenarios, PP 30-47.

Xiaomei, Y. and Ronqing, L.Q.Y. (1999): Change Detection Based on Remote Sensing Information Model and its Application to Coastal Line of Yellow River Delta -Earth Observation Center, NASDA, China.

Zimmerman, E.W. (1951): World Resources and Industries, Harper and New York.

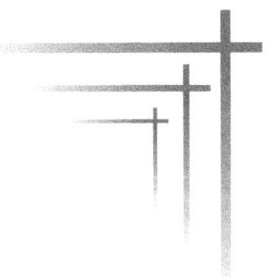

CHAPTER - II
THEORETICAL BACKGROUND

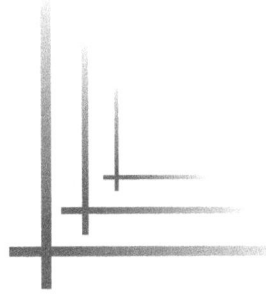

The foregoing chapter was introductory that gave literature review, statement of the problem, objectives and methodology etc. The therotical bacground of the study has been presented here in this chapter.

2.1 CONCEPT OF NATURAL RESOURCES

In common parlance resource stands for available assets. But it has deeper meaning. The prefix re with source is indicative both of function and process. Thus, the word "resource, does not refer to a thing or a substance, but to a function that a thing or a substance may perform or to an operation in which it may take part, namely the function or operation of attending a given end such as satisfying a want" (Zimmerman, 1951). Natural resources (economically referred to as land or raw materials) occur naturally within environments that exist relatively undisturbed by mankind, in a natural form. A natural resource is often characterized by amounts of biodiversity existent in various ecosystems. Natural resources are derived from the environment. This is currently restricted to the environment of Earth yet the theoretical possibility remains of extracting them from outside the planet, such as the asteroid belt. Many of them are essential for our survival while others are used for satisfying our wants. Natural resources may be further classified in different ways. Some examples of natural resources are air, wind and atmosphere, plants, animals, forestry, soils, water, Sun etc. (NRSA, 2007, now NRSC).

Natural resources are natural capital converted to commodity inputs to infrastructural capital processes. They include soil, timber, oil, minerals, and other goods taken more or less from the Earth. Both extraction of the basic resource and refining it into a purer, directly usable form, (e.g. metals, refined oils) are generally considered natural-resource activities, even though the latter may not necessarily occur near the former. A nation's natural resources often determine its wealth and status in the world. The developed nations are those, which is less dependent on natural resources for wealth, due to their greater reliance on infrastructural capital for production. However, some see a resource curse whereby easily obtainable natural resources could actually hurt the prospects of a national economy by fostering political corruption. Political corruption can negatively impact the national economy because time is spent giving bribes or other economically unproductive acts instead of the generation of generative economic activity. There also tends to be concentrations of ownership over specific plots of land that have proven to yield natural resource. The classification of resources shown through **figure 2.1.**

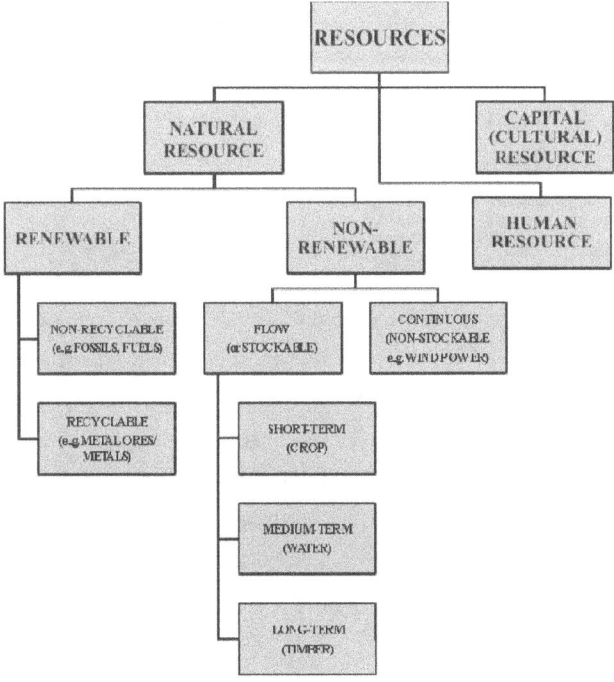

Fig. 2.1

In recent years, the depletion of natural capital and attempts to move to sustainable development has been a major focus of development agencies. This is of particular concern in rainforest regions, which hold most of the Earth's natural biodiversity irreplaceable genetic natural capital. Conservation of natural resources is the major focus of natural capitalism, environmentalism, the ecology movement and green parties. Some view this depletion as a major source of social unrest and conflicts in developing nations. The natural resources, essential for human development and survival are getting depleted and destroyed at an alarming rate. The growing human population and their requirements for food and energy resulted in overexploitation of natural resources. This indiscriminate use of our resources will definitely lead to complete depletion of our natural resources in the nearest future (Rodriguez and Sachs, 1999).

2.2 CLASSIFICATION OF NATURAL RESOURCES

Natural resourcs are mostly classified into renewable and non-renewable resources.

2.2.1 Renewable Resources

Renewable resources are generally living resources (fish, reindeer, coffee, and forests etc.), which can restock (renew) themselves if they are not over-harvested but used sustainably. Once renewable resources are consumed at a rate that exceeds their natural rate of replacement, the standing stock will diminish and eventually run out. The rate of sustainable use of a renewable resource is determined by the replacement rate and amount of standing stock of that particular resource. Non-living renewable natural resources include soil and water. Flow renewable resources are very much like renewable resources; only they do not need regeneration, unlike renewable resources. Flow renewable resources include renewable energy sources such as the following renewable power sources: solar, geothermal, biomass, landfill gas, tides and wind. Resources can also be classified on the basis of their origin as biotic and abiotic. Biotic resources are derived from living organisms. Abiotic resources are derived from the non-living world (e.g. land, water, and air). Mineral and power resources are also abiotic resources some of which are derived from nature.

2.2.2 Non-renewable Resources

A non-renewable resource is a natural resource that exists in a fixed amount that cannot be re-made, re-grown or regenerated as fast as it is consumed and used up. Some non-renewable resources can be renewable but take an extremely long time to renew e.g. fossil fuels, take millions of years to form and so are not practically considered renewable. Many environmentalists proposed to tax on consumption of non renewable resources.

2.3 Concept of Forest Resource

All the forests of the world are not alike. They differ from place to place depending upon certain factors like physiognomy, biomass, life forms, floristic, stratification, phenology and distribution. Physiognomy refers to the physical appearance of the plant communities. Stratification shows the different layers of the various plants. Phenology refers to the seasonal changes in the development of foliage, flowering, fruiting etc. Phenological changes are controlled by species, locality and climatic conditions. The time of development of new foliage varies from species to species; it depends upon locality and climatic factors. Since the physical factors are not homogeneous in nature hence there exists heterogeneity in plant communities on the global as well as micro or regional level (Dwivedi, 1993). Forests are the most important resource of our globe. Naturally formed forests are found in such parts of globe where the factors of plant growth have been ideal for several centuries. For this reason, they are mostly found in regions of high rainfall and regions of high altitude. The usefulness of forests is spread to commercial exploitation of forest for timber and other products, maintenance of birds and wildlife, maintenance of ecological balance, prevention of soil erosion, etc.

In different parts of world, there are evergreen forests (in equatorial belt and parts of tropics), deciduous forests (in parts of tropics and temperate regions), coniferous forests (in temperate and arctic regions) scrub forest (in dry regions). Each one of them is different in respect of composition of species, atmospheric features of its location, density of plants and type of wildlife it has harboured. Forest resources are most important resources of any country useful in maintaining ecological balance, providing fire wood, providing raw materials to many industries, providing protection to wild animals and to conserve the soils. There are mainly four types of forests in India, evergreen forests are found in tropical tract of the country mainly concentrated in heavy rainfall hilly tracts of our country. Deciduous forests are found in plateau region of northern India and parts of central India. Coniferous forests are restricted to Himalayan and Sub-Himalayan regions. Scrub forests are distributed in all parts of the country. Retaining the forest area in tact without encroachment for agricultural and urban uses is one of the greatest challenges faced by the country. Per capita land availability getting reduced every year due to high population growth, forest area is lost at an alarming rate every year mainly to be converted into agricultural land (MoEF, 1999).

Like all other nations, India has its own resource of crops. Although crop resource is dynamic in the sense that many crops and their varieties are lost or added during the course of time, a clear recognition can be made about crops serving as an important national resource. Climatic conditions in the world vary from cold (polar areas) to hot (tropical climates with heavy rainfall) to very dry and hot deserts (that are able to maintain only a minimal amount of plant life). Climatic factors (mainly temperature, precipitation or water availability) determine the distribution of the major forms of natural vegetation in the world. The world vegetation zones are called biomes meaning that major vegetation types extend over large areas in different parts of the world. The significant diversity of forest resources or forest ecosystems is due their presence in a wide range of environments. This extends from the hot and humid tropics to the margins of the deserts and the arctic tundra communities. Within each vegetation zone there is a wide variety of local climatic, topographic and soil conditions. Certain tree species adapt to different ecological circumstances. Human influences can also impact the forest ecosystem, largely in the form of the introduction of exotic species and by the creation of forest plantations. This can often be very productive but is considered to be a simplified ecosystem.

2.4 FOREST PRODUCTS

Forests constitute one of the major natural resource of India. They produce a large variety of woods, which are used as fuel, timber and industrial raw material. They also provide many more things out of which bamboos, canes, herbs, drugs, lac, grasses, leaves, oils, etc. are important. Accordingly the forest products of India are classified into two categories *viz.,* major products and minor products.

2.4.1 Major Forest Products

Major forest products consist of timber, small wood and fuelwood including charcoal. Indian forests produce about 5,000 species of wood, of which about 450 are commercially valuable. Both hard and soft woods are obtained from Indian forests. Hard woods include important species such as teak, mahogany, logwood, iron-wood, ebony, sal, etc. These woods are used for furniture, wagons, tools, etc. Soft woods include deodar, poplar, pine, fir, cedar, balsam, etc. They are light, strong, fairly durable and easy to work and as such are very useful for constructional timbers. They also provide useful raw materials for making paper pulp. It is interesting to note that 70 per cent of hard wood is burnt as fuel and only 30 percent is used in industries while 70 percent of the soft wood is used in industries and only 30 percent is burnt as fuel. Forests meet about 40 percent of energy needs of the country including more than 80 percent of the rural energy requirements. The current production of wood is about 52 million cubic metres out of which about 40 million cubic metres or nearly

77 percent of total is used as fuel wood. The current production of wood is too short of our present demand, which is increasing at an alarming rate. It has been estimated that our annual fuel wood requirements by 2010 will be 350 million cubic metres. The increase in demand for industrial wood will be much more keeping in view the large-scale industrialisation in the country (SFR, 2009).

2.4.2 Minor Forest Products

Minor forest products include all products obtainable from the forests other than wood and thus comprise products of vegetable and animal origin. Some of the important forest products of minor nature are, grasses, bamboos and canes, oils, gums and resins etc.

2.4.3 Indirect Uses of Forests

Apart from their direct uses described above, the forests are of immense use to man indirectly also. They prevent soil erosion, regulate the flow of rivers and reduce the frequency and intensity of floods, check the spread of deserts, add to soil fertility and improve the extremes of climate. Forests play a significant role in the prevention and control of soil erosion by water and wind, control floods during rainy seasons, checks on spread of deserts, increase of soil fertility etc.

2.5 FOREST CONSERVATION

Forests comprise a unique gift of nature to man and constitute one of the prized assets of a nation. They play a significant role in the national economy of a primarily agricultural and developing country like India. The agricultural and industrial progress of the country is not only stabilized but also accelerated by a proper conservation and utilization of forest resource. As mentioned earlier, the uses of forests, both direct and indirect, are so large that they are aptly termed as an index of prosperity of a nation. Keeping in view the benefits, which derive from the forests, it is of utmost importance that strong steps should be taken to conserve forests. The increased demand for forests products has led to increasing destruction and degradation of the forests which is causing heavy erosion of top soil, erratic rainfall and frequent devastating floods. In short, depletion of forests has a chain reaction in eco-system. Though it is a renewable resource, it takes its own time to regenerate (NRSA, 2007, now NRSC). Forest conservation does not mean the denial of use, but rather the proper use without causing any adverse effect on economy or environment. But any scheme of conservation of forests on a piecemeal basis will not solve the problem. Conservation of forests is a national problem and should be tackled as such. There should be perfect coordination between the forest department and other departments for an effective conservation of forests. People's participation in any forest conservation is of vital importance. *Van Mahotsava* was launched in 1950 to make people aware of the importance of planting trees. *Chipko movement* is a living example of general public awareness about forests. Forests have to be developed and worked for obtaining various raw materials and for providing an effective means of flood control, soil erosion, for regulating the flow of water in streams and for conserving moisture in the soil. Therefore, a carefully co-ordinated scientific policy for conservation of forests should be the first step in any scheme of national planning of the country. While contribution of forests to the nation's economy, apart from their vital role in environment, can never be underestimated, the investment in forestry sector has been rather low (Marcot, 1992).

The Forest (Conservation) Act, 1980 enacted to check indiscriminate deforestation diversion of forest lands was amended in 1988 to make it more stringent by prescribing punishment for violations. Guidelines have been prepared for working plans. Some salient features are: (i) working plans should be up-to-date and stress conservation; (ii) preliminary working plan should have multi-disciplinary approach; (iii) tribal rights and concessions should be highlighted along with control mechanism; (iv) grazing should be studied in detail and specific prescriptions should cover fodder propagation and (v) shifting cultivation and encroachments need to be controlled. Conservation of forest is certainly a necessity that requires to be addressed as a priority. For the survival of human beings, a holistic approach is required to be adopted as regards protection of the plant kingdom as well as the wildlife with regard to the peaceful and mutually beneficial co-existence of all. Coming back to India, there are legislations galore to deal with the situation by way of wildlife protection, forest conservation, etc. No doubt, the aims and object of such legislations are in tune with the call of the hour. Deforestation in a rampant manner by various elements has depleted the forest areas forcing the wild life to lesser domains and as a result of this rampage of human habitat by wild a very large mammals and other animals have become a regular phenomenon in certain areas. Forests are being encroached by people who have been displaced from their original habitat for various reasons like construction of huge electricity generating dams, ethnic clashes, floods caused by breach of river embankments and dams etc. (Barbier, 1987).

2.6 FOREST POLICIES IN INDIA

The first scientific forest policy in India was formulated more than a century ago. The charter of Indian Forestry issued by the Government of India in 1855 heralded the beginning of forest conservancy in India. This charter, based on the recommendations of Dr. Mc Clelland, the then Superintendent of Forest, imposed certain restrictions on unchecked exploitation of forests by private agencies. Forests were entrusted to the care of Forest Department. They were surveyed and scientifically mapped. Subsequently, Dr. Voelcker's report on 'Improvement of Indian Agriculture' 1893 included a special chapter on forestry. The first National Forest Policy of India published in 1894 was based on Dr. Voelcker's recommendations (Kant and Cooke, 1999). The Policy classified forests with reference to their primary functions, as preservation forests, commercial forests, minor forests and pasturelands. While preservation forests included those forests essential for environmental purposes, the commercial forests were earmarked for exploitation for timber and other forest produce. Minor forests and pasture lands satisfied the needs of local people for fuel wood, small timber and fodder. Dietrich Brandis, the father of Indian Forestry had a lot of foresight in addressing the needs of the local forest dwellers during the preparation of first Indian Forest Policy 1894. The main objective of forest management was to promote the general well being of the country. The influence of forest over the physical and climatic conditions of the country was quite well recognized. However, realization of maximum revenue from commercial areas was given priority (Negi, 1986). Permanent cultivation was given preference as a land use over forestry; the policy did not make any provision for management of private forests. Due emphasis was not placed on wild life management, catchment area management, forestry research, education and training. These gaps were set right to some extent in the National Forest Policy in 1952 after India's independence. The National Forest Policy of 1952 was formulated on the basis of the following paramount needs, viz.

- The need for evolving a system of balanced and complementary land use.

- The need for checking denudation of mountains and regions on which the perennial water system depended, whose basins constitute the fertile core of the country.

- The need to check erosion along the riverbanks.

- The need to establish tree lands wherever possible.

- The need to ensure progressively increasing supplies of fuel wood, small timber and fodder.

The National Forest Policy of 1952, classified forests into protection forests, national forests, village forests and tree lands. The protection forests were meant for maintaining physical and climatic conditions, similar to preservation forests of the earlier policy. Unlike the earlier policy, tree lands outside the scope of forest management were given special delivery through *Van Mahotsava* programmes. Neighbouring areas were given preferential claim over forests and their produce but not at the cost of National interest. According to the policy agricultural requirements had preferential claim over the forests. It also recommended regulation of rights and restriction of the privileges of users depending upon the value and importance of forests. Therefore, the National Forest Policy of 1952 emphasized complementary land use under which each type of land is allotted that form of use in which it would produce the most and deteriorate the least.

In order to discourage uncontrolled grazing, several measures were suggested in the policy such as rotational grazing, prohibition of grazing in regeneration areas and rationalization of grazing fees, exclusion of goats and creation of fodder reserves. The Indian National Forest Policy of 1988 gave conservation orientation and a human face to forestry (Anon, 1988). The Policy emphasized the protective role of forests in maintaining ecological balance and environmental stability. The basic objectives that should govern the National forest policy have been enlisted as follows:

- Maintenance of environmental stability through preservation and, where necessary, restoration of the ecological balance that has been adversely disturbed by serious depletion of the forests of the country.

- Conserving the natural heritage of the country by preserving the remaining natural forests with the vast variety of flora and fauna, which represent the remarkable biological diversity and genetic resources of the country.

- Checking soil erosion and denudation in the catchment areas of rivers, lakes and reservoirs in the interest of soil and water conservation, for mitigating floods and droughts.

- Checking the extension of sand dunes in the desert areas of Rajasthan and along the coastal tracts.

- Increasing substantially the forest/tree cover in the country through massive afforestation and social forestry programmes, especially on all denuded, degraded and unproductive lands.

- Meeting the requirements for fuelwood, fodder, minor forest produce and small timber of the rural and tribal populations.

- Increasing the productivity of forests to meet essential national needs.

- Encouraging efficient utilization of forest produce and maximizing substitution of wood.

- Creating a massive people's movement with the involvement of women, for achieving these objectives and to minimize pressure on existing forests.

The National Forest Policies (NFPs) are supposed to be revised to strengthen the management, administration and development of the forestry sector, mainly based on the reduction in the forestland area, which is primarily due to forest clearance for agriculture expansion, for livelihood improvement, collection of non-timber forest products, timber extraction, but all these reasons also establish the linkages between local people and the political factors. Within the NFP designs the strategies for forest cover improvement, mainly forest protection and conservation, natural regeneration, forest plantations including high-quality timber plantations with active participation of the private sector is also discussed. Some state governments implement programs suggested by the NFP or other policy prescriptions, at times even without considering whether those programs are suitable for that particular state, or even when there is no proper justification for the implementation of such programs given the climatic, edaphic and vegetative conditions. The target of 33 percent of forest cover should not be an absolute figure, but should be flexible depending on the situational and contextual aspects of the forest resources. A more meaningful parameter would be able to assess the quality of the forest state, its density, its regeneration potential, and trends in the resource quantities and values. Similarly, for forestry schemes meant for poverty alleviation and or improving livelihood of rural population, it is not the extent of forest area that matters to the village communities, but rather a sustainable yield of a variety of forest products with different values, including non-timber forest produce is what is desirable.

The review of the NFP by Indian Institute of Forest Management (IIFM, 2001) has suggested to resolve, protect and improve the environment and forests of the country by initiating key programs including forest protection and afforestation, forest fire control measures, treatment of drought prone areas, strengthening of infrastructure, wildlife conservation, pollution control measures and implementation of environment law. But much of these activities is not justified or well integrated within the forest policy cycle. A new or revised NFP proposal **(Fig. 2.2)** is required to combine the top-down and bottom-up approaches for NFP implementation for a 20-25 years strategy on the concept of forestry development. The National Forestry Programme at every 10 years can oversee the development and transformation of the suggested activities at various management levels; this will ensure a systematic strategy relationship among various management levels and also coordinate mechanisms and procedures for conflict resolution. The National Action Plan (NAP) every 5 years should be integrated with other national strategy partners and collects information, and evaluate rapidly changing areas. Monitoring the critical forest area loss with the use of satellite remote sensing and GIS would assist in this regard. The revised scheme can continue with the creation of a geospatial database system for all natural resource areas, so that a reliable and goal-oriented NFP can be developed (IIFM, 2001).

2.7 STATE FOREST POLICY

Himachal Pradesh, predominantly a mountainous state in the Western Himalayas, has a geographical area of 55,673 sq.km. The altitude of the state varies from 350 metre to 6975 metre above the mean sea level. The state has three distinct regions, viz. the Shiwaliks with altitudes upto 1,500 metre, Middle Himalayan region between 1,500m to 3000m and the Himadris higher than 3,000m. The recorded forest area of the state is 37,033 km.sq. which is 66.52 percent of its geographical area.

Reserved forests constitute 5.13 percent, protected forests 89.46 percent and unclassed forests 5.41 percent of the recorded forest area (SFR, 2011).

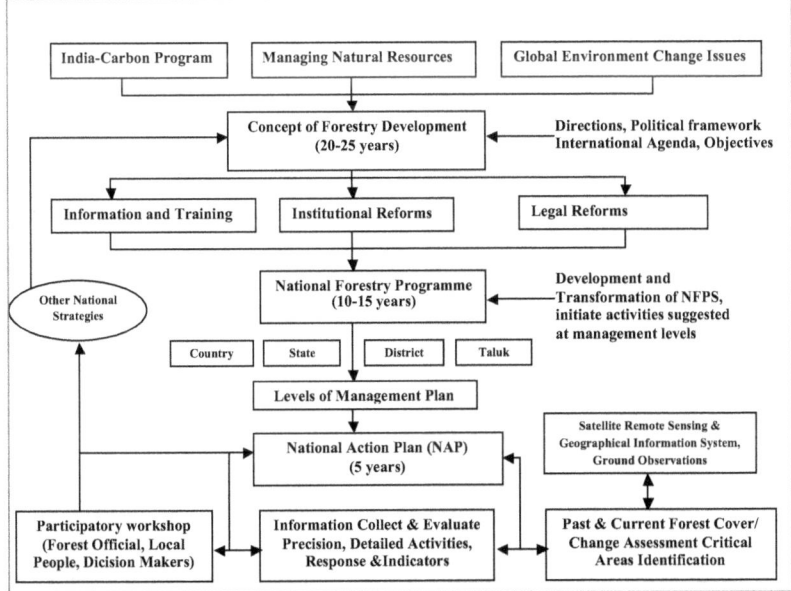

Source: Adopted from Kouplevastskaya (2006)

Fig. 2.2 Concept and design of science-based NFP, as a combination of top-down and bottom-up interdisciplinary approaches for forest policy cycle.

The 1988 National forest policy dictates that environmental conservation of forests and the meeting of subsistence needs of forest-dependent communities should take precedence over commercial production. This is generally supported in Himachal Pradesh; there is a widespread feeling amongst senior government officials in the state that the primary purpose of forests should be for conservation and sustainable use. The state aims to bring 50 percent of the area under tree cover, besides meeting all local requirements. At present there is no formal state forest policy document in Himachal Pradesh. However, the recent Forest Sector Review (FSR) enabled substantial progress to be made towards the development of a new forest policy (Rabindranath and Joshi, 2006). It was a strategic process designed to enable informed debate on key issues affecting the forest sector and represented an opportunity to redefine the state's priorities for the sector. It provided a basis of information and consensus on which to build future policies and strategies for the forest sector. The FSR sets out current thinking on policy for HP's forest sector, which is in line with the national policy climate, and lays the foundation for HP to develop a new forest policy (Himachal Pradesh Forest Sector Review, 2000). Summary of recommendations of the Himachal Pradesh Forest Sector Review Recognition of four key principles:

- There are multiple forest values which sustain local livelihoods and economic growth.

- There are multiple forest stakeholders involved in the forest sector, from those dependent on forests for subsistence to state, national and international stakeholders.

- Changing conditions: whether economic, environmental, social or institutional, conditions are changing rapidly.

- The need for a lead agency to coordinate the transition to SFM.

2.7.1 Himachal Pradesh Forest Policy, 2005

The new Himachal Pradesh Forest Policy of 2005 states "sustainable forest management" as its chief goal and lists the following principles as its priorities: sustainable development, integration of natural resource management, decentralized governance, gender equity, and that forest policy should be more of a process-enabling it to be reviewed, adapted and revised as needed (GoHP, 2005). In its objectives the policy lists the following:

- Conserving and improving the natural resource base (faunal, floral and biodiversity) through effective management based on watershed principles.

- Conservation and management through sustainability and good forest practices - economically, socially and environmentally.

- Providing livelihood security to the forest dependent poor through forest goods and services.

- Participatory approach involving integration and involvement of all stakeholders.

- Meeting forest sector national obligations with regard to policies, laws, international agreements and covenants.

- Forest sector capacity building through research, training, extension, education and awareness.

- Adaptability, monitoring, review, and revision as needed.

- Appropriate land use.

2.7.2 Community Forestry in Himachal Pradesh

Community involvement emerged as a new paradigm in forest management in the nineteen eighties. Participation of local people began to be seen as the solution to rising deforestation rates, as the impact of spontaneous local joint-management initiatives became evident (Poffenberger 1994; Lynch and Talbot, 1995). Policies and projects in India and internationally began incorporating a social component, and people's participation became an essential aspect of forestry projects. National policy also reflected these changes. "The National Forest Policy, 1988, envisages people's involvement in the development and protection of forests. The requirements of fuel-wood, fodder and small timber such as house-building material, of the tribals and other villagers living in and near the forests, are to be treated as first charge on forest produce" (GoI, 1990). Five community forestry projects were functioning in Himachal Pradesh in 2001: DFID Himachal Pradesh Forestry Project (or Joint Forest Management / JFM), Indo- German Changar Project (IGCP), Integrated Watershed Development Project (IWDP), Great Himalayan National Park (GHNP) Eco-development Project (EDP), and Sanjhi Van Yojana (SVY). Each of these projects created village committees to manage forest areas. Institutions and past practices such as the forest settlements, *rakha* system, *devban,* and forest cooperatives are worthy precursors of current community forestry programs. Unlike many other regions of India, forest settlements in Himachal Pradesh have recognized several local rights. The *rakha* is represented by a forest guard with dual accountability to both the state and local communities. *Devban* or sacred groves illustrate a system integrating local belief systems with natural resource management. Forest cooperatives in Kangra district are exemplars of decentralized forest management with local involvement and support.

CONCLUSION

Forests are vital natural resource base that has played a very important role in the economic development of any country. They not only provide the valuable products like timber, fuel wood, fodder, and raw material for many industries and medicinal herbs for the people but also plays critical role in maintaining the ecological balance and keeping the environment free from pollution. Management and sustainable development of forests at the national and state level to determine their conservation and preservation of reverent policies are critical and analytical components. An overview of physio-cultural setting of the study area shall be discussed in next the chapter.

REFERENCES

Anon, (1988): National Forest Policy 1988, Ministry of Environment and Forests, Government of India.

Barbier, E. (1987): The Concept of Sustainable Economic Development, *Environmental Conservation,* 14(2): PP 101-110.

Bahuguna, V. K. (2002): Forest Policy Initiatives in India over the Last Few Years, in the Proceedings of the Forest Policy Workshop, 22-24 January, 2002, Kuala Lumpur.

Bhati, J.P. (1979): Development Strategies in Himachal Pradesh, Mountain Farming Systems Series No. 6, Kathmandu: ICIMOD.

Borgoyary, M., Saigal, S. and Peters, N. (2005): Participatory Forest Management in India: A Review of Policies and Implementation, Working Paper No.1 Overseas Development Group, University of East Anglia, Norwich, UK.

Buchy, M. (1995): The British Colonial Forest Policies in South India, A Maladapted Policy, in Yvon Chatelin and Christophe Bonneuil (ed) Nature and Environment, Orstom Editions, Paris.

Byron, N. and Arnold, J.E.M. (1999): What Futures for the People of the Tropical Forests, *World Development,* 27(5): PP 789-805.

Damayanti, M.O.E. et al. (2004): Institutional Aspect of Protected Areas and Local People's Involvement into the Management: A Case Study in Periyal Tiger Reserve, Kerala, India, Presented at the Conference of Japan Forestry Economic Society, Tsukuba, 21, November, 2004.

Dwivedi, A.P. (1980): Forestry in India, Dehradun, India, Jugal Kishore and Company, New Delhi.

Dwivedi, A.P. (1993): A Text Book of Silviculture, International Book Distributors, Dehradun, PP 975.

Hoskins, M. (1990): The Contribution of Forestry to Food Security, *Unasylva,* 42: PP 160.

Falconer, J. (1990): Hungry Season Food from the Forests, *Unasylva,* 41: PP 160.

FAO (2006): Global Forest Resources Assessment 2005, Rome, Italy.

Forest Research Institute (1953): British Commonwealth Forest Terminology, Part I-Silviculture, Protection, Mensuration and Management, Together with Allied Subjects, *the Empire Forestry Association,* London, UK.

Gadgil, M. and Guha, R. (1995): Ecology and Equity, the Use and Abuse of Nature in Contemporary India, Penguin Books India, New Delhi.

GoHP (2005): Forest Sector Policy and Strategy, Himachal Pradesh Forest Department, Government of Himachal Pradesh, Shimla.

GoI (Government of India) (1952): National Forest Policy (NFP) 1952, New Delhi.

Guha, R. (1983): Forestry in British and Post-British India: A Historical Analysis, *Economic and Political Weekly,* 29 October and 5-12 November.

Guha, R. (1989): The Unquiet Woods, Oxford University Press, New Delhi, India.

Guha, R. and Gadgil, M. (1989): State Forests and Social Conflicts in British India: Past and Present, *Journal of Historical Studies,* 123: PP 143-177.

Helms, J.A. (1998): The Dictionary of Forestry, *Society of American Foresters,* Bethesda.

IIFM (2001): Tree Resources Outside Forest in India, SFM Series, *Indian Institute of Forest Management (IIFM),* Bhopal, India.

Kant, S. and Cooke, R. (1999): Cultivating Peace: Conflict and Collaboration in Natural Resource Management in D. Buckles (ed.), Cultivating Peace: Conflict and Collaboration in Natural Resource Management, Washington, IDRC-World Bank Institute, PP 81-97.

Kaushal K.K., Melkani, V.K. and Kala, J.C. (2005): Sustainable Poverty Alleviation through a Forestry Project in India, *International Journal of Sustainable Development and World Ecology,* 12: PP 1-6.

Kouplevastskaya, I. (2006): The National Forest Programme as an Element of Forest Policy Reform: findings from Kyrgyzstan, *Unasylva,* 225(57): PP 15-22.

Kumar, S. (2002): Does Participation in Common Pool Resource Management Help the Poor: A Social Cost-Benefit Analysis of Joint Forest Management in Jharkhand, India, *World Development*, 30(5): PP 763-782.

Lund, H.G. (1999): A 'Forest' by any Other Name, *Environmental Science and Policy*, 2(2): PP 125-133.

Lynch, O. and Talbott, K. (1995): Balancing Acts: Community-Based Forest Management and National Law in Asia and Pacific, Washington, D.C.: World Resources Institute, Washington.

Malhotra, K.C. and Poffenberger, M. (1989): Forest Regeneration through Community Protection: The West Bengal Experience, Proceedings of the Working Group Meeting on Forest Protection Committees, Calcutta, June 21-22.

Marcot, B.G. (1992): Conservation of Indian Forest, *Conservation Biology*, 6(1): PP 12-16.

Mathur, A.S. and Sachdeva, A.R. (2003): Towards an Economic Approach to Sustainable Development, Planning Commission, GoI, New Delhi.

MoEF (1999): National Forestry Action Programme, India, Government of India, New Delhi.

NRSA (2007) Natural Resources Census: National Land Use and Land Cover Mapping Using Multi-Temporal a WIFS Data, Project Report, Publication No. NRSA/LULC/1:250 K/2007-1, *National Remote Sensing Agency*, Hyderabad, India.

Negi, S.S. (1986): Forest Policy and Five Year Plan, in a Hand Book of Forestry, IBH, Dehradun, PP 102-120.

Pari, B. (1998): The Persistence of Population in Indian Forest Policy, *The Journal of Peasant Studies*, 5: PP 96-123.

Planning Commission (2008): Eleventh Five Year Plan 2007-2012, Planning Commission, Government of India, New Delhi.

Poffenberger, M. (1994): The Resurgence of Community Forest Management in Eastern India. In Natural Connections: Perspectives in Community-Based Conservation, Ed. R. Jeffery, Washington, D.C. Island Press.

Prasad, R. and Kant, S. (2003): Institutions, Forest Management and Sustainable Human Development-Experiences from India, *Journal of Environment, Development and Sustainability*, 5: PP 353-367.

Pretzsch, J. (2003): Forest Related Rural Livelihood Strategies in National and Global Development, *The International Conference on Rural Livelihoods, Forests and Biodiversity,* Bonn, Germany.

Rabindranath, N.H. and N.V. Joshi (2006): Impact of Climate Change on Forests in India, *Current Science,* 90(3): PP 354-361.

Rawat, V., Singh, D. and Kumar, P. (2003): Climate Change and Its Impact on Forest Biodiversity, *Indian Forester,* 129(6): PP 787-798.

Rodriguez, F. and Sachs, J.D (1999): Why Do Resource-Abundant Economies Grow More Slowly, *Journal of Economic Growth,* 4: PP 277-303.

Sarin, M. (1993): From Conflicts to Collaboration: Local Institutions to Joint Forest Management, Society for Promotion of Wasteland Development, Ford Foundation New Delhi, India.

Schlich, W. (1912): Forests and Rainfall, *Nature,* 89: PP 662-664.

Shah, S.A. (1996): Status of Indian Forestry, Wasteland News, PP 14-31.

Sharma, D.D. (2005): Forests Economy and Environment, Kilaso Books, New Delhi.

State Forest Report (SFR) (2005): Forest Survey of India, Ministry of Forest and Environment, Dehradun, PP 214.

State Forest Report (SFR) (2009): Forest Survey of India, Ministry of Forest and Environment, Dehradun, PP 199.

State Forest Report (SFR) (2011): Forest Survey of India, Ministry of Forest and Environment, Dehradun, PP 139-142.

Tucker, R.P. (1982): The Forests of the Western Himalayas: The Legacy of British Colonial Administration, *Journal of Forest History,* PP 112-123.

Zimmerman, E.W. (1951): World Resources and Industries, Harper and New York.

CHAPTER - III
STUDY AREA-
A PHYSIO-CULTURAL SETTING

The previous chapter lays down the theoretical background of the study discussing various concepts pertaining to resources especially the forest resources and discuss the policy issues. The present chapter elaborates the geographic characteristics of the study area under the physio-cultural setting.

3.1 PHYSICAL ASPECTS

The study area of Renuka forest division is situated in the state of Himachal Pradesh which was formed after independence by including a number of erstwhile princely states of Chamba, Mandi, Suket, Sirmour, Bushahr and other smaller states. Later in 1966, when the state of Haryana was carved out from Panjab, the hills area Kangra, Lahaul and Spiti and Shimla of Panjab were added to the state of Himachal Pradesh as they were socially, culturally and geographically contiguous to it. Himachal is situated in the heart of the Western Himalayas. It was called Dev Bhumi (the Abode of Gods) by the ancients. Physiographically the territory can be divided into three zones-Outer Himalayas or the Shiwalik, Inner Himalayas or mid-mountains and the Greater Himalaya or Alpine zone.

The Sirmour district falls in the South-Eastern parts of Himachal Pradesh which constitutes a part of Southern Himachal Pradesh of Himalayian region. The district is further sub-divided into four sub micro regions of Upper Sirmaur Forested Region, Cis-Giri Region, Sirmaur Shiwalik and Kiardun valley. Shimla district bounds it in the North, Solan district in the North-West, state of Haryana in the West and South while state of Uttarakhand makes its eastern boundary. The whole territory of the district with the exception of the Kiarda Dun is mountainous. Its main stream, the Giri, enters the district at its northern most point. It first runs from north-east to south-east and forms the boundary between Sirmaur and Solan. Turning sharply to south-east it runs through Sirmaur district and divides it into two almost equal halves- the Giri-war or cis Giri territory, south-west of the river, and the Giri-par or trans-Giri tract, north-west of it. The residents of these two tracts differ considerably in their habits, manners, and customs. The trans-Giri tract comprises the mountainous country with deep valleys lying between ranges of varying elevation culminating in the majestic Churdhar and the Giri River (District Gazetteer, 1996).

The Sirmour is one of the twelfth districts of Himachal Pradesh with an area of 2825 Sq.Kms. The district lies in the outer Himalayan ranges commonly known as

Shiwaliks between 77°01′12″ and 77°49′40″ East longitude and 30°22′30″ and 31°01′20″ North latitude. The district is predominantly mountainous. There are six tehsils (Nahan, Renuka, Shillai, Paonta Sahib, Pachhad and Rajgarh) and five blocks (Nahan, Paonta Sahib, Pachhad, Sangrah and Shillai) in Sirmour District. The population of Himachal Pradesh as per the 2011 census is 68,56,509 and that of Sirmour District is 5,30,164. The district is primarily rural and some 4,10,765 people live in villages here. The prevailing sex ratio in the rural areas of Sirmour district is 915.

The area of present study, Renuka forest division is also situated in Sirmour district. It lies between 30°31′11″ and 30°52′16″north latitudes and 77°17′34″ and 77°47′38″ east longitudes (Working Plan of Forest, 2014). The geographical area of the division is 987 sq. km. Two wild life sanctuaries viz. Churdhar Wild Life Sanctuary and Renuka Wild Life Sanctuary also fall in the study area. There are five forest ranges in Renuka division namely Renuka, Sangrah, Nohra, Shillai and Kafota. The entire tract is mountainous and varies in elevation from 477 metre to 3647 metre above mean sea level. The slopes are generally steep to precipitous with deep rivulets and springs. The entire region of Renuka forest division falls within the catchments of Giri, Sainj and Tons rivers. The Jalal stream and Naitka stream are two important streams, which drains into Giri at Sieun and Khairi respectively.

3.1.1 Relief

The physical shape of the surface of the earth, its mountains and valleys, plains and plateaus, the physical landscape often applied loosely to indicate inequalities or variations in shapes and forms of the earth's surface. The use of the terms 'high relief' and 'low relief' is best restricted to areas which respectively show a great or little variation in altitude (Clark, 2003). The relief of the study area varies from 477 metre to 3647 metre. The lowest point in the study area is Sataun at the Giri river bed and the highest point is Churdhar peak at Shimla Sirmour district boundaries.

3.1.2 Slope

The upward or downward inclination of a natural or artificial surface, a deviation from the perpendicular or horizontal, the degree or nature of such an incline

(Clark, 2003). Slope can be defined as the degree of inclination of the ground surface or of a profile line from the horizontal surface (Strahler, 1956). The study of surface slopes and surface processes has become an important branch of geomorphology since much of the land surface on earth consists of valley slopes. It is possible to delineate on a slope map any area where slope lies above or below critical value or between two critical values. The critical value would depend on vegetative and soil properties and would fluctuate with soil moisture changes. The mountain slopes in the study area are divided into five major slope categories viz. uniform sloping below 15 degree, gently sloping 16-30 degree, moderately sloping 31 to 45 degree, steep sloping 46 to 60 degree and very steep sloping above 60 degree **(Fig. 3.1)**.

3.1.3 Aspect

The direction in which a valley side or slope faces. In deeply cut east-west oriented valleys, the slopes facing the equator receive more sun and are more attractive to settlement than the shaded sides of the valley (Mayhew, 2009).The direction in which a thing faces, particularly applied to slopes in relation to the sun on account of its effect on settlement and plant growth (Clark, 2003). The flat areas in the study area mostly occur alongside the Giri river channels where large terraces have developed as well as at the mountain tops. The slopes with SW-W-NW aspect have minimal coverage in the study area. The slopes with NE-E-SE aspect are predominant coverage **(Fig. 3.2)**.

3.1.4 Geology

A clear comprehension of the dynamics and mechanisms of the various earth processes is an essential prerequisite for understanding, preserving and restoring the natural environment. Geological studies enable to evaluate the capability of the earth for providing water, minerals and other resources, to select appropriate sites for construction of engineering structures and for disposal of wastes, and to identity lands for supporting agricultural, forestry industrial and other economic activities. Geological/lithological maps incorporate a large and varied set of geological data from field observation as well as laboratory measurements, but they are in part subjective because field measurements are always limited by rock exposure, accessibility and special range under which a human being can see.

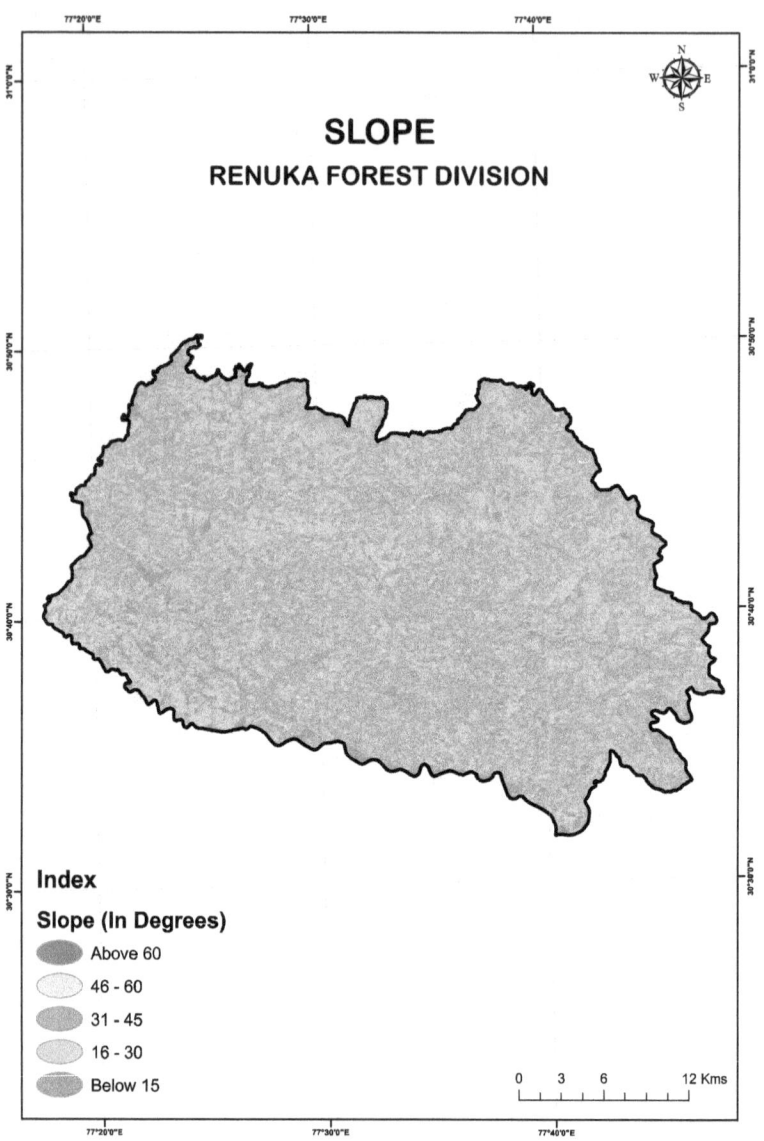

Fig. 3.1

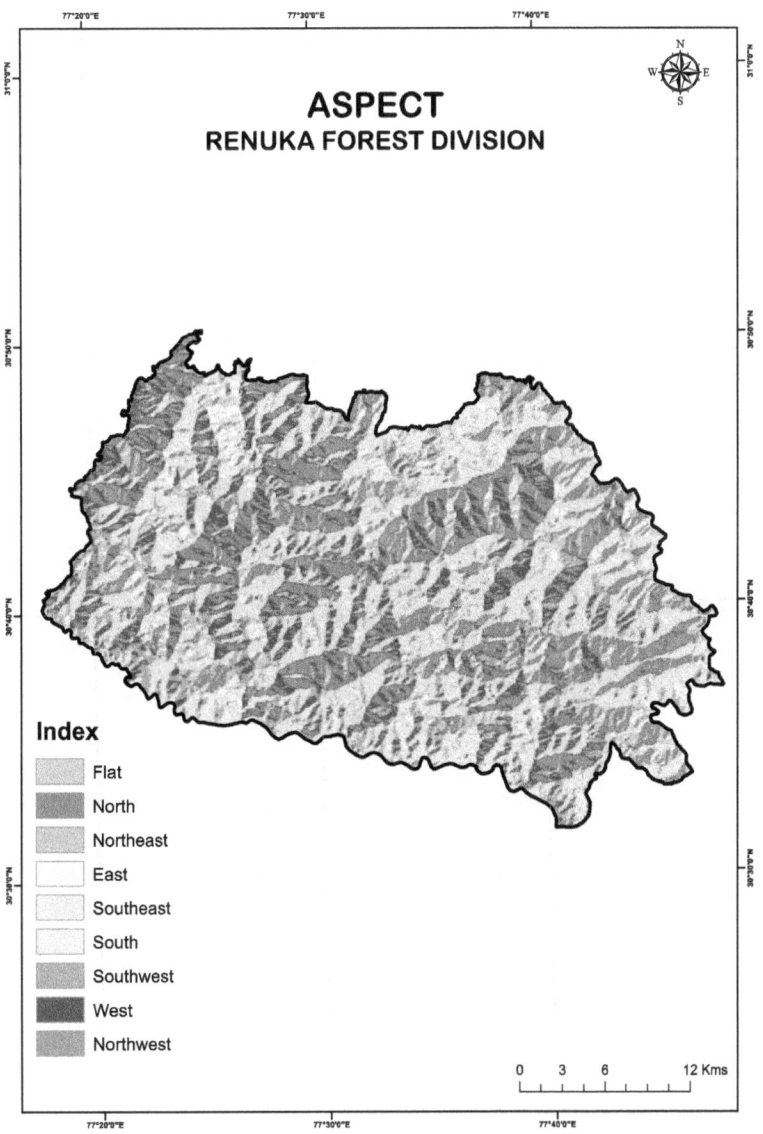

Fig. 3.2

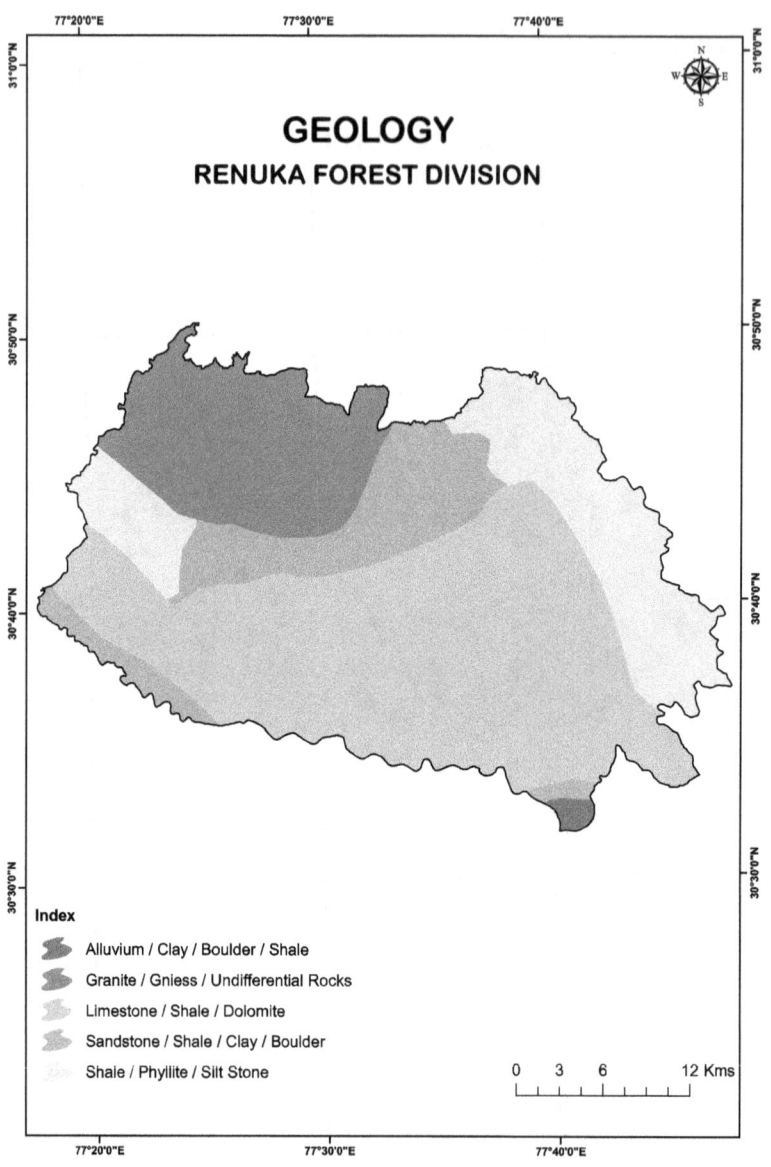

Fig. 3.3

The geology of the study area ranging from Lower Paleozoic to Oligocene is observed in the area. There are quartzite and shale's belonging to Jaunsar group of lower Palaeozoic to upper Proterozoic. The Permian Infra- Krol formation and Triassic Krol-formation along with Jurassic Tal formations, all the three composed of limestone, dolomite and quartzite, are found in the area. These formations are overlain by Subathu Shale, Sandstone and Clays of Eocene to Paleocene age and again by the Shiwalik of Pliocene to Oligocene composed of Clay, Bounders, Shale and Alluvium. Jutogh formations are exposed around Chur Peak in Nohra and Sangrah ranges. Starting from Sangrah in the east it extends to Nohra and then Rajgarh Forest division. The formation consists of carboniferous slates, phyllitic schists wit bands of carboniferous lime stone. It supports deep micaceous clayey loam soils which are fertile and support dense Oak, Fir, Spruce and Deodar forests. Chur Granite consists of coarse grained to prophyritic and gnetiferours granite and genesis. It is exposed along the Chur in Nohra Range (Hayden, 1904).

Shimla group rocks are exposed in North and N-W of this Division along Giri and Tons. Soil from this rock is generally clayey to clayey loam. This formation supports Chil forests. Blaini series occur as thin lenticular outcrops and comprises of carbonaceous, shale, slate, silt, stone, purpled dolomitic limestone and intra bedded red shales **(Fig. 3.3)**. It is exposed in pockets which can be seen above Dadahu. Infra Krol consist of shaly slates with thin carbonaceous quartizite bands and is exposed to south of Bhangal Khad. Deoban formation is exposed in the east of this division along Bhangal and Minus streams (Parimoo, 1983).

3.1.5 Drainage

Drainage pattern of any zone with water surplus has a downstream movement, known as draining (Garner, 1974). The interaction between draining water and physical features results in formation of landforms. A landscape is constituted of a mosaic of draining basins each divided by a watershed and draining to a separate stream system (Briggs and Smithsons, 1985). As drainage basins are catchment's areas for the stream's water, thus, any variation in them affects the stream characteristics. The study area is drained by a number of rivers, rivulets and streams **(Fig. 3.4)**. A little description of them is as under:

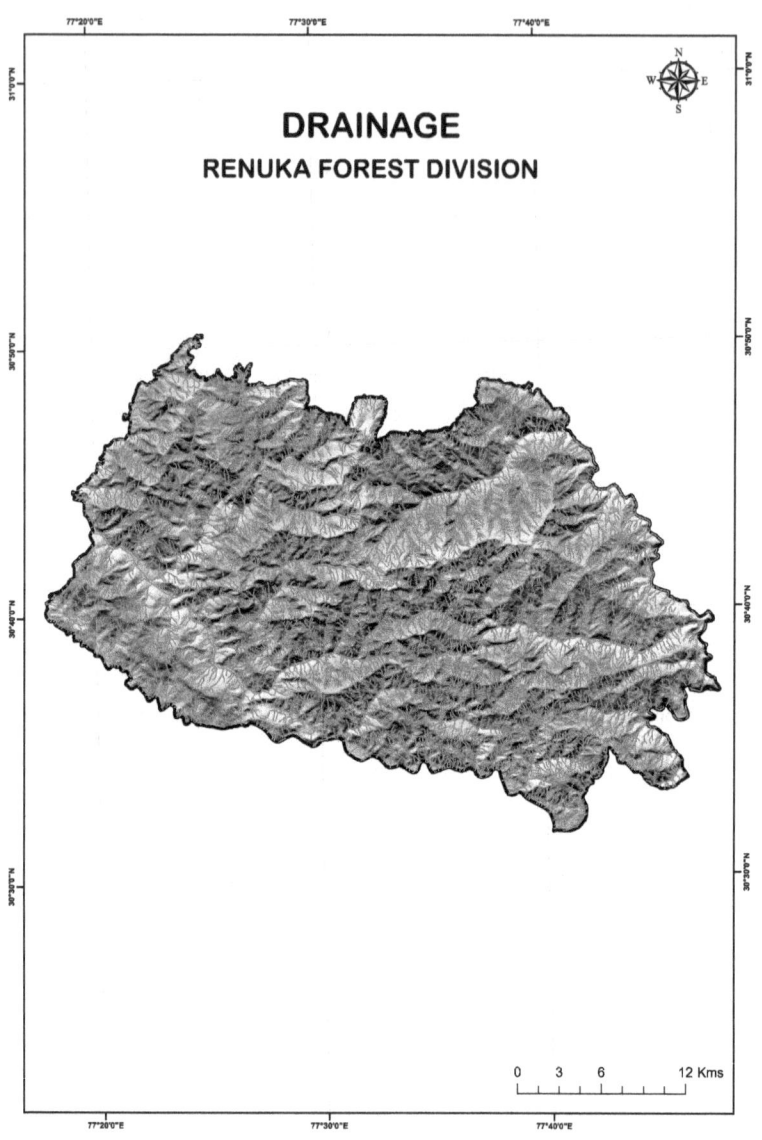

Fig. 3.4

The Giri - The greater portion of the district is drained by the Giri or its tributaries. It takes its rise in the hills of Jubbal and courses through the hills of Kotkhai, parts of Shimla district, and enters the district under study on its south-west side. Its current is swift and water generally muddy while its course is strewn with boulders. The river is of varying width in the Sirmour district, the average being about 122 m. It is, however, for the most part shallow, the depth ranging between 1.2 to 1.5 m, which is fordable. But during the rainy season or sometimes in the summer too, if the rains are excessive and the river is in space, it is difficult to ford it. None of its tributaries is important, except, on its right bank, the Jalal, which joins it at Dadahu below Sati Bagh at the south-eastern extremity of the Saindhar. On its left bank the principal streams are the Nait and Palar, which rise on the southern slopes of the Choor peak. Below Nehi, in the west, rises the Kawal, a stream which first flows westward and then turns north till it falls into the Giri. Lesser tributaries are the Bajhethu, the Perivi, the Kohal and the Joggar streams.

The Tons - The source of this river lies in the Yamnotri Mountains and after coursing through the territories of Jubbal and Jaunsar it enters the district near village Koti, separating it from the Jaunsar area, once a part of the erstwhile princely state of Sirmour. After flowing for about 50km and forming the eastern boundary of the district it joins the Yamuna near Khodar Majri which is trebled in its size after the junction of the two rivers.

The Jalal - This small, shallow, and narrow river rises near village Bani below Nehi in tahsil Pachhad and forms a dividing line between the Sain and the Dharthi. At Dadahu in tahsil Renuka it falls into the Giri River, losing its name.

3.1.6 Soils

Soil is just not an anchorage and nutrient reservoir for plants; it is essentially a product of the plants and dead organic matter, which inhabits it. The weathered rock mantle and the plants both contribute to its formation and are inseparably linked by a continual interchange of materials. Soil is the topmost live layer of the earth's crust formed by the mechanical and chemical weathering of the parent rock and its reaction with the partly decomposed organic matter and the bacteria derived from the earlier generation of organism under the action of climate, (heat, rainfall and wind) and vegetation. It may take over hundred years for a few centimeters of soil to be formed.

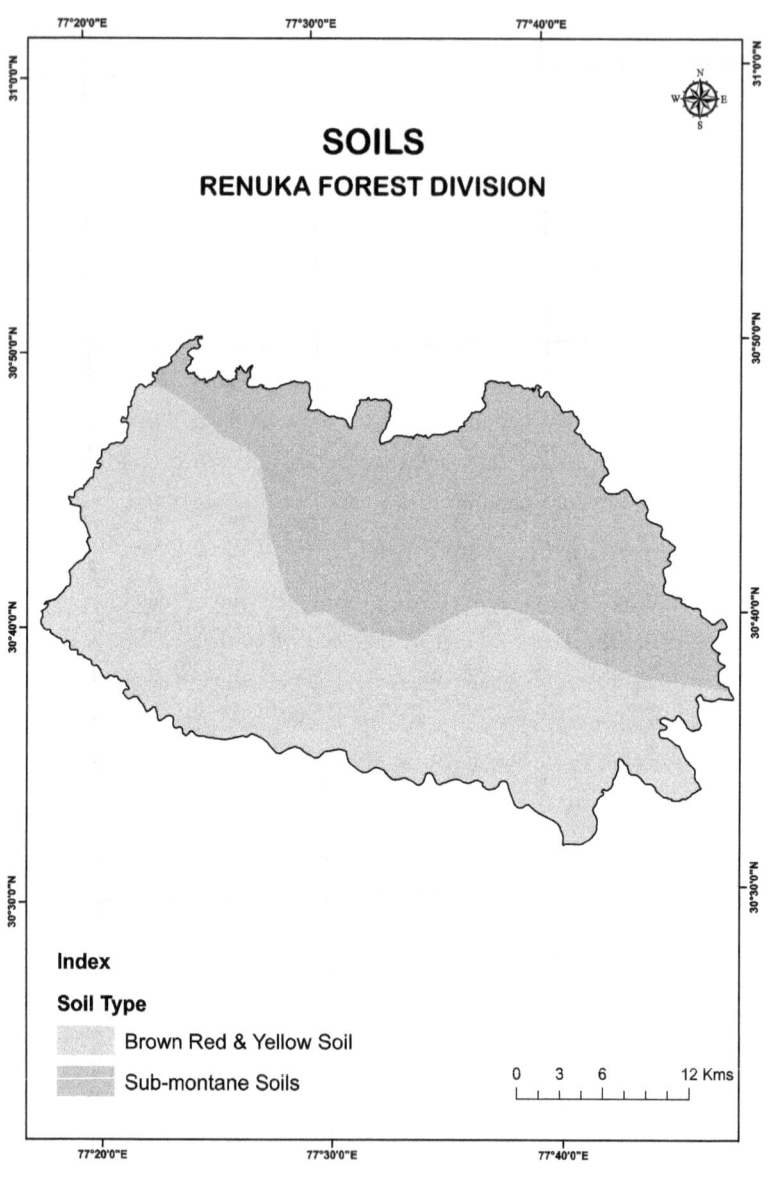

Fig. 3.5

Yet it cannot be said as universally true because in certain climatic conditions particularly in temperate grassland, soil formations are much faster. Vegetation in turn also influences soil. In vertical section a soil shows a succession of natural layers called horizons resulting from the leaching and translocation of material usually from surface downwards.

These differ in colour, texture reaction, consistency and porosity within a homogeneous climatic region. A close correlation exists between soil conditions and forest vegetation. Of the various soil properties those important from the forest point of view are its depth, texture, structure, organic matter content, moisture, PH and nutrient level. Deep soil with a good moisture status, a high nutrient reserve is the most favorable for forest growth. As most forest trees have a high physiologic need for moisture, it is essential that forest soils possess a high moisture contents to meet the demand. A large proportion of clay and organic matter increases the water holding capacity of the soil. Organic matter formed from the decomposition of fallen plant debris is the most important constituent of soil. The leaves, twigs and other parts of tree when decomposed under the influence of heat (temperature) and moisture add to the nutrients value of the forest soil. The scientific classifications of soil recognize 75,000 different kinds of soils. Each is different from other in depth, size arrangement of particle, in mineral compositions, water-holding capacity etc. The profiles are well developed in higher location under dense forests. Lower down these suffer from erosion. Soil is clayey loam, sandy clayey loam and clay. On ridges spur and precipitous slopes the soil is shallow. There are mainly two types of soil in the study area; brown red and yellow and sub montane soil **(Fig. 3.5)**.

3.1.7 Climate

The climate is perhaps the most important factor in any study related to ecology. The climate influences the human life both directly and indirectly. The study of the climate thus becomes and imperative in a study particularly related to the ecology of the forests. Temperatures, moisture, humidity sun light direction of the winds, snowfall etc. all the elements of climate influence the forests right from the germination of seed to the maturity. Not only of an individual tree but a whole complex of vegetation gets influenced by the climatic elements in one way or the other, in the words of Dikshit, "whole the land form and surface analysis requires, and carries the imprint of climatic elements like temperature and its range, the humidity, the dry spell, the number of rainy days and the intensity of rain, the ecological profile of a place hinges most on the bioclimatology of the area. The latter rests more on the retention of soil moisture regularity of rainfall, temperature within specific range, ground water conditions, atmospheric humidity and finally a combination of temperature, rainfall, soil moisture and the level of evapotranspiration, that would ensure year long growth and existence of plant life" (Dikshit, 1991).

The study area possesses a variety of climate varying according to elevation. Temperature experiences both hot in summer as well as severe cold in winter and rainfall occurs maximum in summer and snow in winter in the higher altitude of the region. Summer months are extremely and exceedingly hot in the study area, the hilly areas have a temperate climate though Dharati range is hot. The Trans Giri tract, are cool even in the hot weather (Fig. 3.6). In Trans Giri area, snow falls every winter while in Dharati it falls rarely. The climate of upper part of study area forested region is cool in summer and cold in winters. Maximum parts of this region receive snow fall in winter. In upper area of Cis-Giri region, the climate is cold in winters and mild in summer while lower areas are hot in summer and cool in winter. This region receives sufficient rain falls.

Rainfall is highly variable in the region due to its rugged orography and its geographical position. The rainfall generally increases up to the valleys from southwest to northwest and decreases beyond the highest range over the northern part of the region. The percentage contribution of rainfall is less than the southern parts, perhaps due to decreasing influence of the monsoon. The rainy season usually begins in the middle of June and lasts till the middle of September. A shower or two are received in April and May. The April and May rains may also bring hailstorms. The snowfall on higher ridges begin in December and lasts in March. Churdhar remains snow covered for major part of the year. During monsoon, rains are more active in the month of July and August. About 80 per cent of the rainfall is received by the district during July and August months. After April, the rainfall gradually increases till June and thereafter sharply during July and August. It decreases rapidly after the withdrawal of southwest monsoon in September.

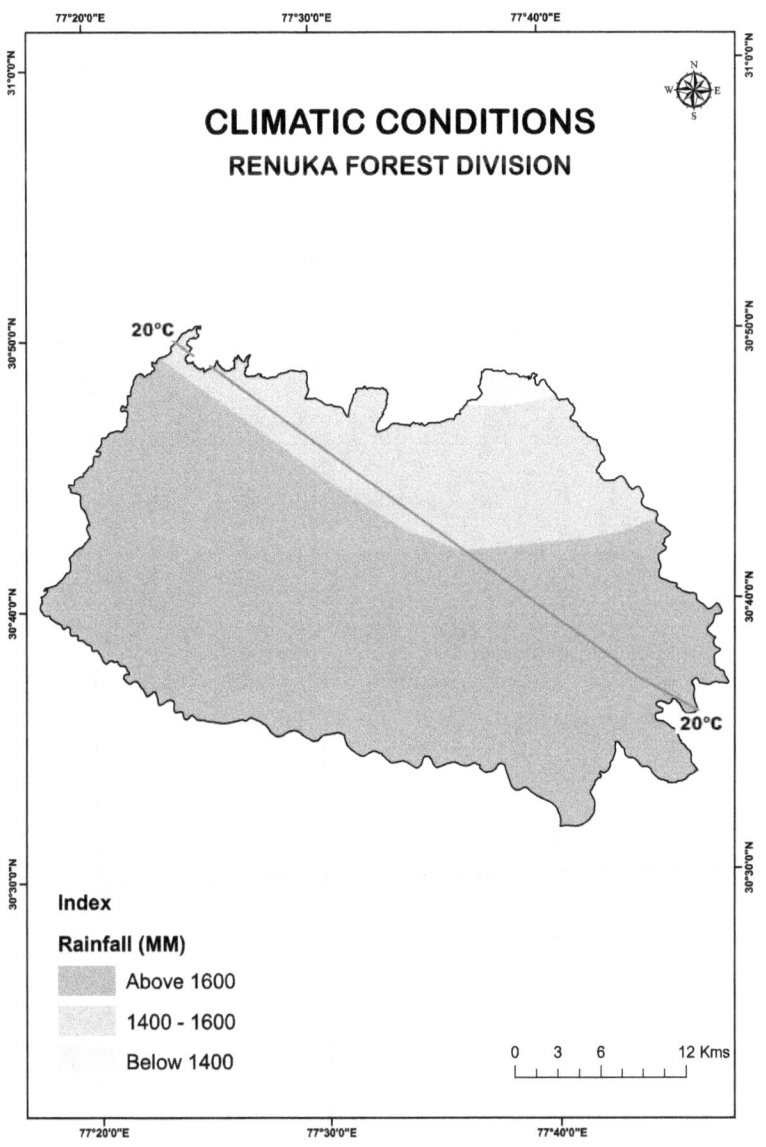

Fig. 3.6

3.1.7.1 Rainfall: The average rainfall data for the last ten years that is from 2001 to 2011 recorded at Renuka 645 (msl) is tabulated in **(table 3.1)** below:

Table: 3.1 Average Rainfall at Renuka 2001-2010

Month	Average Rainfall (mm)	No. of rainy days
January	50.8	4
February	81.3	5
March	51.48	6
April	30.6	4
May	39.25	5
June	112.08	9
July	222.5	12
August	200.17	10
September	116.93	7
October	51.05	3
November	26.82	2
December	119.75	4

Source: Indian Meteorological Department, Shimla

3.1.8 Natural Vegetation

The vegetation of the study area is diversified due to wide range of topography, altitudes, soils and climate. It can be studied under four categories as:

3.1.8.1 Tropical Forests (up to 800m): This type of forests is represented by Sal (*Shorea robusta*) as the dominating species.

3.1.8.2 Montane Subtropical Forests (800-1800m): Pinus roxburghii is the dominating species with other associated species of flowering plants as Berberis lycium, Emblica officinalis, Rubus ellipticus, Terminalia chebula, Toona ciliata, etc.

3.1.8.3 Montane Temperate Forests (1800-3000m): These forests occur in pure as well as mixed forms. Cedrus deodara occurs in pure strands at Haban and Haripur Dhar. At other places, the various species as Pinus wallichiana, Pyrus pashia, Rhododendron arboreum, Quercus semecarpifolia, etc. are present. Common shrubby undergrowth includes those of, Indigofera gerardiana, Rosa macrophylla, Sarcococca saligna, Viburnum cotinifolium, etc.

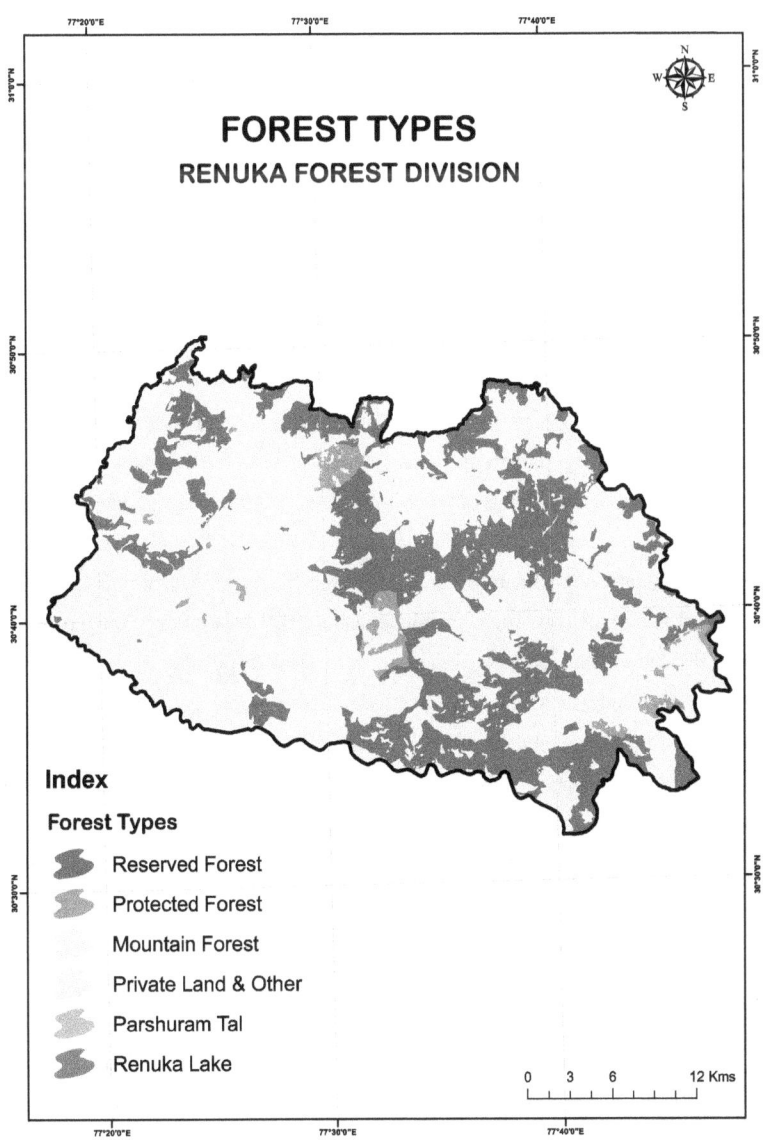

FOREST TYPES
RENUKA FOREST DIVISION

Index

Forest Types

Reserved Forest

Protected Forest

Mountain Forest

Private Land & Other

Parshuram Tal

Renuka Lake

0   3   6   12 Kms

Fig. 3.7

3.1.8.4 West Himalayan Upper Oak-Fir forests (above 3000m): This type of forests is found only in Churdhar areas. Main constituent is Quercus semecarpifolia with scattered trees of Abies pindrow and Picea smithiana, Common shrubs are Prunus cornuta, Rosa sericea, R. webbiana, Sorbus foliolosa, etc.

The recorded forest area of the study area is 549 km.sq. which is 55.62 percent of its geographical area. Reserved forests constitute 89.4 percent, protected forests 10.4 percent and unclassed forests 0.2 percent of the recorded forest area (Fig. 3.7).

3.2 CULTURAL ASPECTS

It is human population from which all other elements are observed and from which they all singly or collectively derive significance and meaning (Trewarth, 1953). Man's knowledge is considered as the mother resource that has the potential for creating the resources. Zelinsky (1966) considers, "resources as progeny of human aspiration, memory, talent and labour applied to relatively inert physical entities". In order to evaluate the human resource in the region, the growth of population, its spatial distribution, density, socio-economic characteristics and occupational patterns have to be studied. And some of these important attributes of human resource with respect to the study area is being described in the following discussions which influence the forest ecology the most. In the study area there is distinctive culture, different from cis-Giri area. They have there pahari culture, joint family system, rigid caste system, diverse marriage and moral system, etc. people believe in collective approaches in the rural area for working in field, social customs, traits, different kinds deeds in the villages. The people are very innocent and hard working in their nature, always busy in their fields and rearing and rendering of domestic animals.

3.2.1 Population Distribution

Demography is the study of population change over time and space and it studies the various determinants of population change and the impact of such changes on socio-economic development of region. The study of population gives an idea about the dispersion of population. The density is one of the parameters for measuring population ratio of the region. This can be measured by arithmetic, agricultural and physiological. The population distribution is studied in terms of population concentration. The distribution of population means the dispersion of population in study region. The differences from place to place in racial and social character of population are studied in social science (James, 2003). From this point of view it is interesting to study the population and their changes in study region. The change in population is not only change in its numbers but also its change in structure, composition and distribution with respect to region and time. The population growth means changes in total population, it may be positive or negative. Population growth is the indicator of economic and social development. The study of measurement of such change, both temporal and spatial and comparative study gives an idea about changing characteristics of population of study region. According to census of India 2011, shows that total population of the Renuka forest division are 1,67,705 (Table 3.2, Fig. 3.8) and density is 169 people per sq. km.

Table: 3.2 Range wise Population Distribution of Renuka Forest Division, 2011

Sr. No.	Name of Range	Total Population
1	Renuka ji	30712
2	Sangrah	40000
3	Nohra	23022
4	Shillai	36905
5	Kafota	37066
Total		1,67,705

Source: Census of India, 2011

3.2.2 Literacy Pattern

Literacy is an indispensable mean for acquiring skills and improving economic and general well being. From demographic point of view, literacy is key variable affecting fertility, mortality and migration. Literacy supports for development and population control too (Singh and Yadav, 2010). According to 2011 children age of 7 years or below are treated illiterate even though they may be going to school and can read and write to same extent. Average literacy rate of Sirmour in 2011 is 78.80 compared to 70.39 of 2001. If things are looked out at gender wise, male and female literacy were 85.61 and 71.36 respectively. For 2001 census, same figures stood at 79.36 and 60.37 in Sirmour district. Total literate in Sirmour District were

362,645 of which male and female were 205,617 and 157,028 respectively. Literacy rate of Renuka forest division is 75 percent (Census of India, 2011). Maximum population is living in the village and mainly engaged in primary activities they believe in agriculture practices and take low education.

3.2.3 Sex Ratio

The human population exhibits certain inherent characteristics in terms of sex composition (Joshi and Tiwari, 2011). The change in sex composition is largely reflecting the underlying socio-economic and cultural patterns of society. The ratio between male and female is called as sex ratio. In India, sex ratio defines the number of females per 1000 males. It is an important social indicator to measure the extent of prevailing equality between males and females and it is an outcome of interplay of sex differentials in mortality, sex selective migration, sex ratio at birth and times, sex differential in population enumeration (Census of India, 2011). Sex ratio affects on economic development, migration, occupation structure, marriage status, fertility, mortality and population growth etc. in any region. The present sex ratio of the study area is 902.

3.2.4 Occupational Structure

There are three basic need of human being, namely, food, shelter and clothes. In order to fulfill this need mankind involves in occupation. The economically active mean the population, which actually takes part or tries to take part in production of goods and services. People working in different sectors are in various occupations. It is one of the factors of economic development in region, 'the economic and social development of persons who are economically active, quality of their work and the regularity of their employment.' (Bhende and Kanitkar, 2006). If more people are engaged in primary activity means that the region is undeveloped, if the more people are engaged in secondary activity means that region is process of developing and if the more people engaged in tertiary activity means the region is developed.

A study of the labour force covers employment, unemployment and underemployment as well as industrial and others activities. Such study provides the base for social and economic development and it is useful for policymakers and planners (Bhende and Kanitkar, 2006).

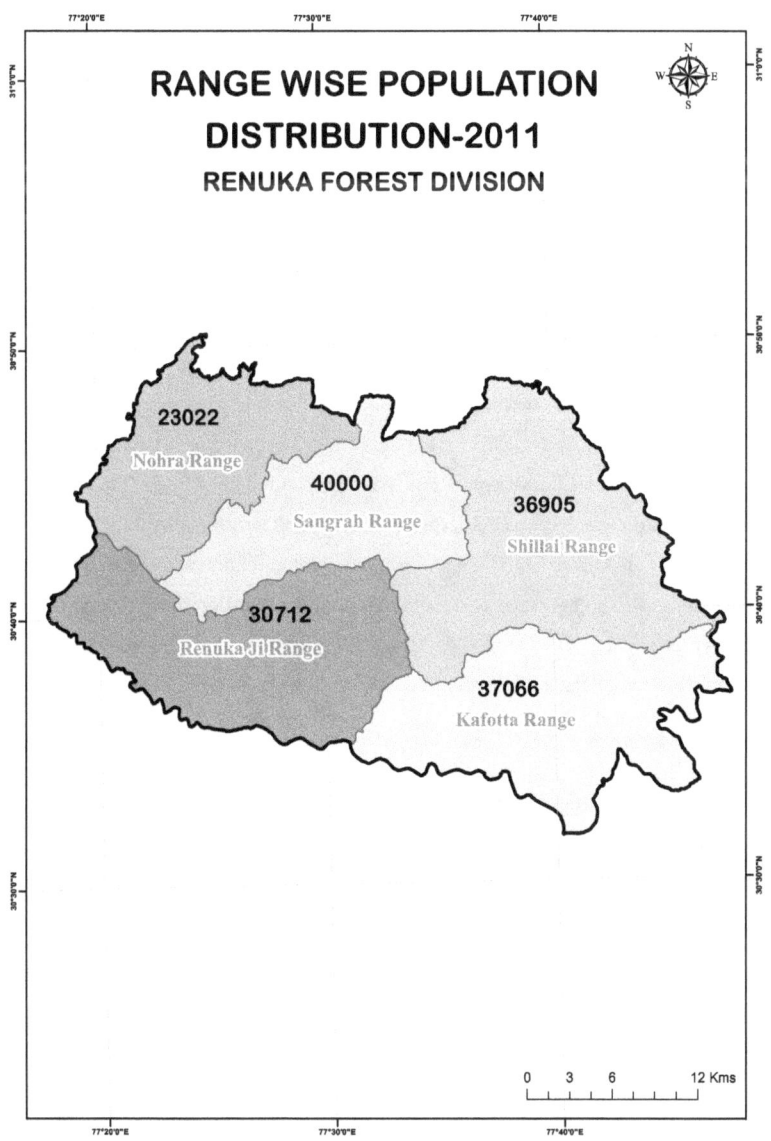

Fig. 3.8

In Renuka forest division, maximum population is living in the village and mainly engaged in primary activities they believe in agriculture practices. Work participation rate is 53 per cent in Renuka forest division.

3.2.5 Other Socio-Economic Activities

The area under rice and wheat recorded increase but the area under maize, pulses and barley decreased in the Renuka forest division. In case of important crops like maize, rice and wheat, the percent of area under these crops witnessed decrease implying some shift in area from these crops to other commercial crops mainly vegetable or commercial crops. The potato, tomato, pea, garlic and ginger are important vegetable and commercial crops both in terms of area and production. The infrastructural facilities available in the study area, over the years there have been marked improvements in transport and communication networks. The health and veterinary institutions have also improved. However, the rural markets and enterprises and industries are still scanty that need special attention. Farmers of the study area are very progressive and are eager to adopt modern agricultural production technologies. They are producing wide ranges of agricultural, horticultural and vegetable crops through adoption of scientific methods.

CONCLUSION

The physical and cultural aspects are very important constituents to determine the existence of forests in any region i.e. aspect, slope, soil, climate and population distribution etc. The forest types, classification of vegetation, forest cover change and people's opinion about forest cover shall be discussed in the next chapter.

REFERENCES

Bhende, A. and Kanitkar, T. (2006): Principal of Population Studies, Himalaya Publishing House, Mumbai.

Briggs, D.J. and Smithsons, P. (1985): Fundamental of Physical Geography, Hutchinson, London, PP 11-14.

Buck, J.L. (1951): The Indian Rural Problem, Vora and Company Publisher Ltd., Bombay.

Chandana, R.C. (2014): Population of Geography, Kalyani Publishers, New Delhi.

Chandana, R.C. and Sidhu, M.S. (1980): Introduction to Population Geography, Kalyani Publishers, New Delhi, PP 17.

Census of India (2011): Census of India, *Ministry of Home Affairs,* Govt. of India.

Clark, A.N. (2003): Dictionay of Geography, Penguin Books.

Demko, G.J. (1970): Population Geography: A Reader, McGraw-Hill Book Co., New York, PP 22.

Dikshit, K.R. (1991): Environment, Forest Ecology and Man in the Western Ghats, Rawat Publication, Jaipur, PP 23.

District Gazetteer (1996): Gazetteer of the Sirmour District, Indus Publishing Company.

Freeman, T.W. (1968): Geography and Planning, Hutchinson University Library, London, PP 74.

Garner, H.F. (1974): The Origin of Landscape: A synthesis of Geomorphology, Oxford university Press, London, PP 27-31.

Ghosh, B.N. (1985): Fundamentals of Population Geography, McGraw-Hill, New York.

Gibbs, J.P. (1961): The Measurement of Change in the Population Size of an Urban Unit' in J.P. Gibbs (ed.) Urban Research Methods, PP 107.

Singh, O.P. and Pandey, D.C. (1986) Development Planning, Theory and Practice, Gyanodaya Prakashan, Nainital, PP 111.

Hayden, H.H. (1904): The Geology of Spiti with parts of Bashar and Rupshu Mem, Geological Survey India, 36(1).

James, K.S. and Sathyanarayana, K.M. (2011): Demographic Change, Age Structure Transition and Ageing in India-Issues and Challenges, *Yojana,* New Delhi, 55, PP 21-27.

James, M.R. (2003): An Introduction to Human Geography, Pearson Education, Inc., Upper Saddle River, NJ.

Joshi, A. and Tiwari, N. (2011): Sex Ratio in India-Embarrassing to be Honest, *Current Science,* 101(8), PP 1006-1008.

Kochhar, P.L. (1967): Plant ecology genetics and Evolution, Atma Ram and Sons, New Delhi, PP 10.

Misra, S.P. (1985): Integrated Rural Area Development and Planning, Ratan Publications, Varanasi (India), PP 41.

Mayhew, S. (2009): Oxford Dictionay of Geography, Oxford University Press.

Parimoo, M.L. (1983): Report on the Base Metal Investigation in District Kullu and Sirmour, H.P. Unpub. Rep. G.S.I.

State Forest Report (SFR) (2011): Forest Survey of India, Ministry of Forest and Environment, PP 139-142.

Singh, J. and Dhillon, S.S. (1987): Agricultural Geography, Tata McGraw- Hill Publishing Co. Ltd., New Delhi, PP 63.

Singh, A.K. and Yadav, K.V. (2010): Population Growth, Causes and Impacts-A Case Study, Uttar Bharat Bhoogal Patrika, Gorakhpur, 40 (4): PP 99-104.

Strahler, A.N. (1956): Quantitative Slope Analysis, *Bulletin of the Geological Society of America,* 67, PP 571-596.

Trewarth, G.T. (1953): A case for Population Geography, *Annals of the Association of American Geographers,* June 1953, PP 71-79.

Waggoner, P.E. (1953): Stem and Root Temperatures, *Phytopathology,* 43: PP 317-318.

Working Plan of Forest (2014): Working Plan for the Forests of Renuka Forest Division, H.P. Govt., Forest Department.

Zelinsky, W. (1966): A Prolongue to Population Geography, Prentice Hall Inc., Englewood Cliff, N.J., PP 103.

CHAPTER - IV

SPATIAL DISTRIBUTION OF FOREST RESOURCES IN RENUKA FOREST DIVISION

The previous chapter was based on physio-cultural setting of the study area. The present chapter gives an overview of the forest types, earlier history of forest of the study area and changes in forest cover of the Renuka forest divison.

4.1 THE FOREST AND THEIR TYPES

All the forests of the world are not alike. They differ from place to place depending upon certain factors like physiognomy, biomass, life forms, floristic, stratification, phenology and distribution. Physiognomy refers to the physical appearance of the plant communities. Stratification shows the different layers of the various plants. Phenology refers to the seasonal changes in the development of foliage, flowering, fruiting etc. The phenological changes are controlled by species, locality and climatic conditions. The time of development of new foliage varies from species to species; it depends upon locality and climatic factors. Since the physical factors are not homogeneous in nature hence there exists heterogeneity in plant communities on the global as well as micro or regional level (Dwivedi, 1993). It is therefore, natural to understand the basic nature of the forests formations, their associations and inter-relationships so as to understand their resource base in its totality in given physical environment. Such a study leads to differentiating one type of vegetation from another depending on factor like climate, altitude locality aspects, etc. A number of attempts made for the classification of vegetation in different part of the world and a brief resume of that has been given here.

4.1.1 Classification of Vegetation

Classification of vegetation can be approached from the study of plant communities, the data of which can be grouped in any of the several categories such as structures and composition etc. The data may break down into successively smaller units on the basis of differences or can be aggregated into successively large units on the basis of similarities. In the one case, the primary units are the plant associations, number of which is assembled into a formation or type, in the other case, a series or formations sub-divided in stages into the constituent associations. An important advance in the study of climate and vegetation was made by Koppen (1931). He distinguished eleven main climatic zones initially on the basis of temperature, rainfall and season. Later, he used primarily the temperature of the hottest and coldest months and the relative rainfall of the wettest and driest months. This classification is however, based more on reactions of vegetation with climate than on metrological data. Thornthwaite (1933) differentiated thermal zones on a basis of temperature efficiency (TE) with numerical values for a TE index obtained by summing the 12 monthly indices each classified in geometrical zones giving 50 types of vegetation.

In 1936, Champion attempted to classify the vegetation of the world on the basis of climatic features but later in 1968 there was a revision in this classification. All the classification of the vegetation of the world are not applicable to India owing to the complexities of the environmental factor including climate, geology, soil forests biota and forest history. "In an ancient country like India where much of the climax vegetation has long been destroyed, biota and edaphic features are considered more important for the present structure, composition and status of various vegetation types" (Champion, 1936). All this not withstanding, Champion's classification of Indian vegetation.

Thornthwaite's (1948) later formula involving the evaluation of potential evapotranspiration may be considered here rather than under bioclimatic methods. Although evapotranspiration naturally depends on the variety, density, life form and stage of growth of vegetation, he uses the concept of potential evapotranspiration, under unlimited supply of moisture the value being independent of the nature of the vegetation. This also is not very satisfactory because the actual climatic characteristics of many stations departing significantly from Thornthwaite's types. Nevertheless, it fits the distribution of vegetation types better than those mentioned earlier. It may, however, be said that the complex nature of the relationships between evaporative processes and plant behavior remain to be worked out and that attempts to picturize the relationships between evaporation and forest types must necessarily be painted with a broad brush (Turner, 1961).

The role of high temperatures in limiting active growth of vegetation has been recognised by Shanbhar (1958). Later in 1960, Puri put forward a classification of Indian vegetation. He considers that there is no justification for separating the northern and southern variations of the major types the differences might be due to biotic or edaphic factors and that vegetation changes occur more in an east-west direction. He also assumes that the characters of the dry deciduous thorn and evergreen types are ascribable to biotic factors and not to climatic differences. This assumption is also not acceptable; although it can at once be accorded that several of the characters in question are generally accentuated by biotic factors and the area occupied by these types has been widely extended by their action. Finally there is a differentiation of subsidiary from primary edaphic types which is difficult to follow, and of biotic types still more difficult to distinguish from the secondary edaphic group.

As far as forest types of Himachal Pradesh are concerned few people have attempted to classify the vegetation type of the state. The classification followed in the working plans of various forest division of the state forest department can form the basis for filling in the details by Champion and Seth. Based on the classification forest types of India by Champion and Seth (1968) and taking some clues from the working plans prepared by the foresters an attempt is being made to identify the forest type of the study area- Renuka forest division in the following discussion.

The aspect and the edaphic factors play the significant role of supporting the vegetation whereas the biotic factors lead to the degeneration stages of the forest. The parts of Nohra, Sangrah and Renuka ranges face the southern aspect which being warmer is dry and the portions facing the northern aspect are moist and support ban, fir, spruce and deodar forests. The Shillai and Kafota ranges are rich in limestone and portions of this edaphic site support Khair, Shisham and mixed deciduous forests with Terminalias and Anogeissus as the main species. Grazing and heavy lopping are the chief biotic factors may to lead the degradation. The type of forests of the Renuka forest division met as per 'A Revised Survey of the Forest Types of India' by Champion and Seth (1968) falls under groups of 5, 9, 12 and details of these forest groups are given as below:

GROUP 5   **TROPICAL DRY DECIDUOUS FORESTS**

SUB GROUP 5 B   **NORTHERN TROPICAL DRY DECIDUOUS FORESTS**

1.5 B/C 1a   *Dry Shiwalik Sal*

2.5 B/C2   *Northern Dry Mixed Deciduous Forests*

GROUP 9   **SUB TROPICAL PINE FORESTS**

9/C 1   *Himalayan Chir Pine Forests*

3.9/C 1a	*Lower or Shiwalik Chir Pine Forests*
4.9/C 1b	*Upper or Himalayan Pine Forests*
GROUP 12	**HIMALAYAN MOIST TEMPERATE FORESTS**
12/C 1	*Lower Western Himalayan Temperate Forests*
5.12/C 1a	*Ban Oak Forests*
6.12/C 1b	*Moru Oak Forests*
7.12/C 1c	*Moist Deodar Forests*
8.12/C 1d	*Western Mixed Coniferous Forests*
9.5/DSI	*Dry Deciduous Scrup Forests*
10.5/IS 2	*Khair-Sissoo Forests*
11.9/C1/DS 1	*Himalayan Sub-tropical Scrub*
12.9/C1/DS 2	*Sub Tropical Euphorbia Scrub*
13.12/C1/DSI	*Oak Scrub*

4.1.2 Type 5 B/C 1a Dry Shiwalik Sal

This type of forests is met within exposed southern slopes of Shiwalik ridge and is exposed to periodical dry spells. The slopes are fairly steep while the soil is shallow and sandy with pockets of clay. Soil is devoid of humus and posses boulders, pebbles or beds of sand stone. Sal is the main species. The proportion of Sal is 60 per cent. The other associates are *Terminalia tomentosa* (Sain), *Anogeissus latifolia* (Chal), *Lannea coromanelica* (Jhingan), *Buchanania lanzan* (Chiroli), *Acacia catechu* (Khair), *Boswellia serrata*. Occasional trees of Chir pine are also met with. The undergrowth consists of *Carissa opaca, Flemingia chapper* (Saipan), *Randia dumetorum, Woodfordia fruticosa* (Dhau), *Colebrookia oppositifolia* etc. grasses are *Euliolopsis binata* (Bhabar) and *Heteropogon contortus* (Lamb). Climbers are *Bauhinia vahlii* (Taur) and *Milletia auriculata*. The natural regeneration is deficient. These types of forests are seen in Janjhli, Sataun etc. in Kafota Range (Kumar et al. 2012).

4.1.3 5B/C2 Northern Dry Mixed Deciduous Forests

This type or forests occurs on the lower slopes draining into Giri, Tons, and lower altitudes of the Division. These forests extend up to 1200 metre from mean sea level. The most common species are *Anogeissus latifolia* (Chal), *Lyonia acidissima* (Beli), *Lannea grandis* (Jhingan), *Aegle marmelos* (Bel), *Flacourtia ramontchi* (Kandai), *Eugenia Jambolana, Mallotus phillipensis* (Kamela), *Terminalia tomentosa* (Sain), *Terminalia chebula* (Harar), *Sygyzium cumini* (Jaman), *Cassia fistula* (Amaltas) and *Ougenia dalbergioides* (Sandan). The undergrowth is moderately dense and generally comprises of *Carrisa opaca* (Karaunda), *Woodfordia fruticosa* (Dhau), *Murraya koenigii* (Gadhelu), *Euphorbia royleana* (Danda Thor), *Adhatoda vesica* (Basauti) and *Colebrookia oppositifolia* (Binda). The common climbers in this area are *Bauhinia vahlii* (Taur), *Pueraria tuberosa* (Sural), *Caesalpinia sepiaria* (Alai) and *Combrelun decandorum* (Ruel). The canopy is light, irregular often broken and formed entirely or deciduous trees. The trees are short and crooked in form. The quality on the whole is poor and the crop is of little value for fuel and fodder. The forests are generally invaded by local people for grass collection, fuelwood and small timber. The thinning of lower storey for fuel and fodder has led to the invasion of weeds like *Lantana camara* (Panjphuli) and *Cassia tora* (Panwar). During summer the forests gives a sparse, stunted and poor appearance whereas in monsoon forests appear lush green with undergrowth and greening of all trees. This type occur mainly in Unger, Thana Kegwa, Charighati, Nehla Gawahi in Renuka Range, Gabbar, Sakhauli, Chandni, Kathar, Dhab-Pipli, Mailani, Sehbara, Manal Salag, Shiva, Sherli Manpur forests in Kafota range and Lagnoo, Mashoor, Gatlog, Kufar Kiara forests in Sangrah Range/ Nohra Ranges (Kumar and Choyal, 2013).

4.1.4 9/C1a Lower or Shivalik Chil Pine Forests

This sub type is confined to Kafota Range and occurs upto 1000 metre elevation on dry southern slopes. The Chil trees are found scattered in pure form or with a scattered deciduous lower storey generally on cool aspects or in depressions. The density is thin and quality poor while the regeneration is absent. The main floristics is Pinus longifolia with *Mallotus phillipensis* (Kamela), *Pyrus pashia* (Kainth), *Emblica officinalis* (Amla). The undergrowth is mainly *Carissa opaca* (Karaunda), *Rubus ellipticus* (Hinsalu), *Myrisne Africana* (Banwan), *Colebrookia oppositifolia* (Binda), *Murraya koenigii* (Gadhelu) and *Adhatoda vesica* (Basauti). This sub type is met in forests Janjhi, Salag of Kafota Range (Kumar et al. 2012).

4.1.5 9/C1b Upper or Himalayan Chil Pine Forests

This type is found from 1200 to 1800 metre elevation over-lapping the Tropical dry deciduous forests at lower elevation and running into temperate forests at higher reaches. This type mainly occurs in Shillai and Kafota Ranges. The chief species is Pinus longifolla. However, at lower elevations along streams in damp and humid depressions and declivitis, *Syzgium cumini* (Jaman), *Lannea grandis* (Jhingan) is found whereas *Quercus leucotrichophora* (Ban), *Rhododendron arboretum* (Buras), *Cedrus deodara* (Deodar), *Pieris ovalifolia* and *Myrica nagi* are found at higher elevation. In Kafota, Sangrah and Shillai ranges there are quite a few examples wherein chil and deodar are growing side by side, Chil occupying spurs and deodar cool springs. This is mainly on northern aspect. Due to heavy grazing and frequent fires tender trees and shrubs have been destroyed and replaced by hardy species such as Flemingia prostrate and Rhus parviflora which are coming up as weeds. At lower levels, however, regeneration is poor due to heavy biotic pressures. The Chil zone is prone to severe soil erosion. Landslips are common wherever road passes through forests. The sub type occurs generally in Charag, Ganu, Cho Bhoghar forests in Renuka Range; Tatiyana, Jamna Pabar forests inn Kafota Range; Balikothi, Koti Bonch, Kharkhan, Loja, Manal, Bandauli forests in Shillai Range and Jamal Nihog, Bhajond forests in Nohra Range (Kumar and Choyal, 2013).

4.1.6 12/C1a Ban Oak Forests

These are the main forests found in the division. They occur at elevation from 1800 to 2300 metre elevation. There are continuous patches of vast areas of these forests. The trees are badly lopped around villages and paraos, and give a short, stunted and open look. However, in the areas away from habitations and paraos, the trees are huge with large crowns. The main species is *Quercus leucotrichophora* (Ban) with mixture of *Rhododendron arboreum* (Buras), *Pieris ovalifolia, Litsea umbrossa, Myrica nagi* and *Cornus capitata*. The undergrowth is generally dense comprising of *Myrisne africana* (Banwan), *Boenninghausenia albiflora* (Pissumar), *Indigofera gerardiana* (Kali Kathi), *Rubus niveus* (Hinsalu), *Desmodium tillaefolium* (Safed Kathi), *Berberis chitria* (Kasmal) and *Prinsepia utilis* (Bhekhra). Among the climber are *Hedrahelix* (Kural), *Smilax parviflora* (Ushwa) and *Vitis trifolia* (Pani vel). Nirgal bamboos also occur in depressions and declivities. The trees are laden with epiphytic ferns and moss. The patch sowing of deodar wherever attempted in oak forests is successful e.g. Ghataun, Tatiyana. This sub type of forests can be seen in the following forests. Renuka Range: Ghatoun, Kafota Range: Tatiyana. Shillai Range: Shri Kyari, Chyali, Bhatnaul, Kota Pab, Khatva, Milla, Jaswi, Lani, Baror, Dabar, Jarwa, Jakandon, Nai Panjore, Tatwa Beyong, Sangrah Range: Dasakana, Taikri, Panjah, Bhaltar, Lajwa, Arat, Ranphua, Uncha Tikker, Nohra Range: Manal, Chokar, Pipli, Bandal, Shilli, Bhangar, Bhangari, Nohra, Bhog, Charna, Ghandoori, Chunvi and Sail (Kumar et al. 2012).

4.1.7 12/C1b Mohru Oak Forests

It extends from 2000 to 2500 metre elevation. The Ban oak forests in lower elevation extends into Mohru oak forests at higher elevation e.g. RF Khatna. A dense forest of Mohru oak near Haripurdhar is a good example. The drier aspects are covered with poor growth of Ban oak. The species is *Quercus himalayana* (Mohru) mixed with *Quercus leucotrichophora* (Ban), *Rhododendron arboretum* (Buras) and scattered trees of *Pinus wallichiana* (Kail). This sub types is seen in following forests: Sangrah Range: Dasakana, Khalandon, Deori-Kharan, Nohra Range: Ghandoori, Sail and Chunvi.

4.1.8 12/C1b Moist Deodar Forests

The forests occur in Kafota, Shillai, and Nohra ranges and cover wide area. These forests occur between 1800 to 2000 metre elevations. The deodar forests are almost pure on exposed ridges. However, in humid rivulets species of *Quercus Leucotrichophora* (Ban) and *Rhododendron arboretum* (Buras) as trees, *Lonicera augustifolia* (Chalu), *Viburum cotonifolium* (Talanj), *Berberis Chitria* (Kasmal), *Daphne cannabina* (Niggs), *Rubus niveus* (Hinsalu), *Prinsepia utilis* (Bhekhra) as shrubs and *Hedera helix, Jasminium officinale* (Ban chameli), *Clematis montana* (Garol), *Rosa moschata* (Bal gulab) are the common climbers. The ground cover consists of *Fragaria indica* (Saipan), *Ranunculus* (Daren) and grasses. The crop is predominantly middle aged. The regeneration is patchy and in many places deficient and absent. The deodar trees are badly lopped for fuel and leaf manure by the locals.

This sub type is seen in the following forests: Kafota Range: Nigali, Tatiyana. Nohra Range: Bandal, Bhangari.

4.1.9 12/Ctd Western Mixed Coniferous Forests

This sub type occurs at elevations ranging from 2400 to 3000 metre mostly in Shillai and Sangrah ranges. The main crop is spruce at lower limits and fir predominates on higher and damper locations. Kharsu oak is the main associate in the under storey. The under growth is moderate and comprises of *Viburnum* (Tanlanj), *Skimmia laureola* (Gurl Patta), *Deutzia corymbosa* (Batti), *Valeriana hardwickii* (Swak), *Ainsliaea aptera* (Aerons rod), ferns and grasses. The tree canopy is moderate, often broken. The trees have tall boles and reach upto 50 metre height. Lightening damage is common occurrence in these forests. The regeneration of spruce is adequate; however, the regeneration of fir is deficient and poor. The saplings of young trees of spruce are often lopped by the Gujjars in the paraos for the roofing of their huts. In midst of oak and fir spruce forests meadows are found at places. These are used by Gujjars for summer grazing. This sub type is met in the following forests: Sangrah Range: Gata Mandwaj, Ghataun, Deori Kharan. Shillai Range: Manal and Luja (Kumar et al. 2012).

4.1.10 5/DS1 Dry Deciduous Scrub

These types of forests are found in the dry deciduous forest zones with poor soil. The main species occurring in these forests are *Lannean grandis* (Jhingan), *Acacia catechu* (Khair), *Aegle marmelos* (Bel), *Carissa opaca* (Karaunda), *Euphoria royleana* (Danda Thor), *Sapium insigne* (Ritha), *Woodfordia floribunda* (Banksha) etc. These forests are met within Renuka, Kafota, Shillai and Nohra Ranges.

4.1.11 5/1S2 Khair Sissoo Forests

These forests occur mainly along the banks of Giri River in Renuka and Kafota ranges. The plantations of Khair along river beds at Sataun have shown good results. Shisham has also started coming up in this riverine succession. The undergrowth comprises of *Zizyphus jujube* (Kath ber), *Murraya koenigii* (Gadhelu), *Adhatoda vesica* (Basauti) etc.

4.1.12 9/C1/DSI Himalayan sub-Tropical Forest

These areas are found mainly in Kafota and Shillai ranges. Such areas occur on southern slopes in chil zone where the soil is shallow. Chil has almost disappeared due to adverse biotic factors such as fellings, grazings and fuelwood collection etc. thereby leaving residual scrub forests of *Sapium insigne* (Ritha), *Euphoria royleana* (Danda Thor), *Rhus parviflora* (Tung) etc.

4.1.13 9/C1/DS2 Sub Tropical Euphoria Scrub

Euphorbia royleana is found in degraded rock outcrops in Chill zone. Due to fast and easy vegetative propagation and its succulent nature it is extending into many forest areas as well. These scrubs are seen on banks of Tons river in Shillai range; Steep slopes on banks of Giri in Renuka and Nohra ranges.

4.1.14 12/C1/DS1 Oak Scrub

The Oak forests on the southern aspect with shallow soil and low moisture show stunted and slow growth. These are also prone to heavy lopping and trees and tend to become scrub. The usual associates are *Pyrus pashia* (Kainth) and *Pieris ovalifolia*. The under growth is of *Rubus ellipticus* (Hinsalu), *Prinsepia utilis* (Bhekhra) and *Berberis* (Kasmal). This degradation stage is met with mainly in Sangrah and Shillai Ranges.

4.2 EARLY HISTORY OF THE FORESTS IN RENUKA FOREST DIVISION

No authentic record regarding the type of vegetation and forest management in the the study area are available. Like other parts of the country, forests of Sirmaur District had richness, vastness and covered places for wildlife hunting. Mr. John Northen says in his book, 'Guide to Masuri, 1984, "Nearly the whole of the dominions of the Raja of Sirmaur is one vast forest, the open valleys, a dense jungle of high grass, and the consequences is, that instead of thousands upon thousands of happily and contented villagers, the land is given upto the beasts of the field and the birds of the air. It is useless to dwell on the short-sightedness of a policy so manifestly opposed to every principle of political economy. Timber might pay a contractor; it never paid a Nation. If population is the wealth of a country, it is useless to ask it to feed on timber. The earning of a prolific population pays the most of the State in hundred ways." By this description, it is clear that the tract was occupied by dense forests inhabited by wild animals like Elephant, Tiger, Panther, Sambar, Chittal, Barking deer etc. With the passage of time, it seems that the policy of the Government changed and clearance of woods to bring more and more land under agriculture continued unabated. Land revenue being the main source of revenue to the State, agriculturists like bahaties and sainies were brought from Hoshiarpur areas of Punjab to clear the forests of Dun valley. During the regime of Raja Shamsher Prakash, at one stage, it was felt that there has been extreme destruction of forests leading to high floods (Working Plan of Forest, 2014).

4.2.1 Jiva Ram's Working Plan (1904-05 to 1924-25)

A working plan for the management of the forests of Pachhad Tehsil was prepared by Babu Jiva Ram, the then Divisional Forest Officer under the guidance of Mr. G.G. Minnikin, Deputy Conservator of Forests and was subsequently revised by the officiating Deputy Conservator of forests, Nahan. This working plan dealt with the areas of Pachhad Tehsil. However, prescriptions for Oak forests falling in Renuka Tehsil were given.

4.2.2 Sewal's Provisional Scheme (1933-34 to 1942-43)

It dealt with all Government owned forests of Pachhad and Renuka Tehsils. Separate sets of proposals were made on the basis of crop for each tehsil. Deodar-Kail and Chil forest of Pachhad tehsil were grouped into the Deodar Working Circle and the Chil Working Circle for commercial exploitation on scientific lines. However, Renuka tehsil forests (Chandpur and Haripur Ranges) tehsil are mostly blank. Fir, Spruce and Oak were confined to a limited area at higher elevation. Chil and Deodar were found in small patches here and there. 'Sewal's scheme prescribed that these forests were to be protected and fellings were to be done only to meet the legitimate requirements of the local population. Blanks were to be fenced and artificially regenerated.

4.2.3 Period from 1942 to 1961

Though sewal's scheme expired in 1942-43 yet the main provisions of this scheme continued to be followed till 1961, when Mukerjee's plan came into force.

Earlier, fellings had been done in accordance with annual felling programme approved by Chief Conservator of Forests, H.P.

4.2.4 Mukerjee's Working Plan (1961-62 to 1975-76)

The two economically important species viz. Deodar and Chil were brought under Shelter wood management during Mukerjee's Plan which came into operation in the year 1961. The following working circle was created :

The Deodar Working Circle

The Chil Working Circle

The Selection Working Circle

The Protection Working Circle

Mukerjee's Plan:

- To preserve and improve forest cover so as to prevent denudation and erosion in hill slopes and ensure even flow of water in streams.

- To provide for the bonafide domestic and agricultural requirement of the local populace for grass, grazing, timber, fuelwood and minor forest produce.

- To bring about a normal distribution of age classes and establish normal regeneration.

- To extend the area under important timber yielding species on suitable lands.

- Consistent with above to obtain the highest possible sustainable yield of timber and other forest produce in perpetuity.

4.2.5 O.P. Sharma's Plan (1976-77 to 1990-91)

There was much difference between the general objectives of management in Mukerjee's Plan and the Sharma's Plan. The general objects of managements in case of O.P. Sharma's Plan were as under:

- To conserve and improve the existing forest cover so as to prevent denudation and soil erosion and thus help sustain and even improve the environment.

- To provide for bonafide domestic and agricultural requirements of the local population for grass, grazing, firewood and minor forest produce.

- To provide timber for fruit packing cases to the orchardists and right holders.

- To bring the growing stock of important economic species to a condition as near to that of a normal forest as possible.

- To restock the degraded forests, particularly the mushterqua forests with valuable species.

- To regulate and control both local migratory grazing as far as possible.

- Consistent with the above, to obtain the highest possible sustained yield of timber, pulpwood, resin and other forest produce.

4.3 FOREST COVER OF THE RENUKA FOREST DIVISION

The assessment of changes in forest cover, between 1972 and 2011 has been analyzed with the help of remote sensing and geographic information system (GIS) **(Table 4.1)**, in the Renuka forest division of Sirmour district. The trend of forest cover changes over the time span of 39 years has been examined. Forests are the green blankets that are naturally protecting the hill environment and preserving the natural resources. The recent researches show that the overwhelming population pressure, practicing of unscientific agricultural methods and the lack of awareness about the importance of forests among the populace in general are the prime causes for deforestation and degradation of forests. The rates of depletion, reason for the deterioration and remedial measures to restore it are the essential factors to assess the forest cover in any terrain. The inventory of forest resources and forest cover assessment and change detection in the rugged topography or hill sector is not an easy task and it is a time-consuming process. This can be made easier only through the high spectral, spatial and temporal resolution qualities of remote sensing techniques. Indeed, the precise database pertaining to forest cover information is an imperative input of formulating various management plans and also remote sensing technology can be effectively utilized for change detection and monitoring activities (Jessica et al. 2001). According to Macleod and Congalton (1998), in general, remote sensing considers following four aspects of change detection (a) detect the changes, (b) identify the nature of change, (c) measure the aerial extent of change and (d) assess the spatial pattern of change. Earlier, many researchers have carried out the change analysis through visual or digital interpretation. Forest cover change detection has been done, through visual interpretation of satellite data by Unni et al. (1985), Roy et al. (1991a,b), Sukumar (1991), Porwal and Pant (1989), Kushwaha (1990).

However, the following researchers Jessica et al. (2001), Pradhan and Awang (2008), Sakthive et al. (2010) Bharti et al. (2011), Hansen (2013) and Stibig et al. (2014) have done the forest cover change detection through computer assisted digital image processing (DIP) techniques. The basic principle of change detection through remote sensing is that the changes in spectral signatures commensurate with the change in land cover. The detailed procedure is to superimpose two period maps to find the change (Jessica et al. 2001). Moreover, the process of change detection is premised on the ability to measure temporal impacts (Sabins, 1987). According to Singh (1989), change detection is the process of identifying differences in the state of an object or phenomenon by observing it in different times (multi-temporal variations). It is evident that change detection can be precisely calculated using GIS technology and because of its high volume spatial and a spatial data handling capability. It enables to do overlay process with two or multi vector layers under single umbrella (Bhaduri et al. 2009). Some of the researchers have identified that the increase in vegetation cover has resulted in increased rainfall (Sharma, 2001; Dengiz et al. 2009) and decrease in forest cover has direct relationship with socioeconomic status and marginal worker force (Murali et al. 2002). Forests are a dynamic feature on the land surface. As true for other covers, forests too change in time and space. The changes may be positive i.e., re-growth, plantations etc., or negative such as degradation and depletion of forests due to population pressure other unscientific practices etc.

Table: 4.1 Forest Cover of the Renuka Forest Division from 1972-2011

	Forest Cover 1972		Forest Cover 1989		Forest Cover 2001		Forest Cover 2011	
Forest Cover	Area in Sq.km.	% age	Area in Sq.km.	% age	Area in Sq.km.	% age	Area in Sq.km.	% age
Forest	610	61.80	634	64.24	558	56.53	549	55.62
Agriculture	178	18.03	115	11.65	102	10.33	190	19.25
Open Land/Grass Land/Shrub Land	190	19.25	230	23.30	320	32.42	242	24.52
Water Body	9	0.91	8	0.81	7	0.71	6	0.61
Total Area in Sqkm.	987	100.00	987	100.00	987	100.00	987	100.00

Source: Data calculated by author from satellite imageries of Landsat MSS, TM, ETM+ and Resource Sat-II

4.3.1 Forest Cover of the Renuka Forest Division in 1972

The spatial distribution of forests cover in 1972 is vividly shown in **figure 4.1**. Forests occupied by 610 sq.km, which is about 61 percent of the total study area. The area under agriculture and open land/grass land/shrub land in the study area were 178 and 190 sq.km respectively i.e. these categories constitutes about 18 percent and 19 percent of the study area respectively. Water bodies occupied 9 sq.km, which is about 0.9 of the study area.

4.3.2 Forest Cover of the Renuka Forest Division in 1989

The spatial distribution of forests cover in 1989 is shown in **figure 4.2**. In the year 1989, the forests occupied 634 sq.km, which are about 64 percent of the study area. The area under agriculture and open land/grass land/shrub land in the study area were 115 and 230 sq.km respectively i.e. these categories occupied about 11 percent and 23 percent of the study area, respectively. Water bodies composed of 8 sq.km, which is about 0.8 of the study area.

4.3.3 Forest Cover of the Renuka Forest Division in 2001

The spatial distribution of forests cover in 2001 is shown in **figure 4.3**. In the year 2001, the forests occupied 558 sq.km, which is about 56 percent of the total study area. The area under agriculture and open land/grass land/shrub land in the study area were 102 and 320 sq.km respectively. These categories occupy about 10 percent and 32 percent of the areal extent of the study area respectively. Water bodies constituted 7 sq.km, which is about 0.7 percent of the study area.

4.3.4 Forest Cover of the Renuka Forest Division in 2011

The spatial distribution of forests cover in 2011 is shown in **figure 4.4**. In the year 2011, the forests occupied 549 sq.km, which are about 55 percent of the study area. The area under agriculture and open land/grass land/shrub land in the study area were 190 and 242 sq.km respectively i.e. these categories occupied about 19 percent and 24 percent of the study area, respectively. Water bodies constituted 6 sq.km, which is about 0.6 percent of the study area.

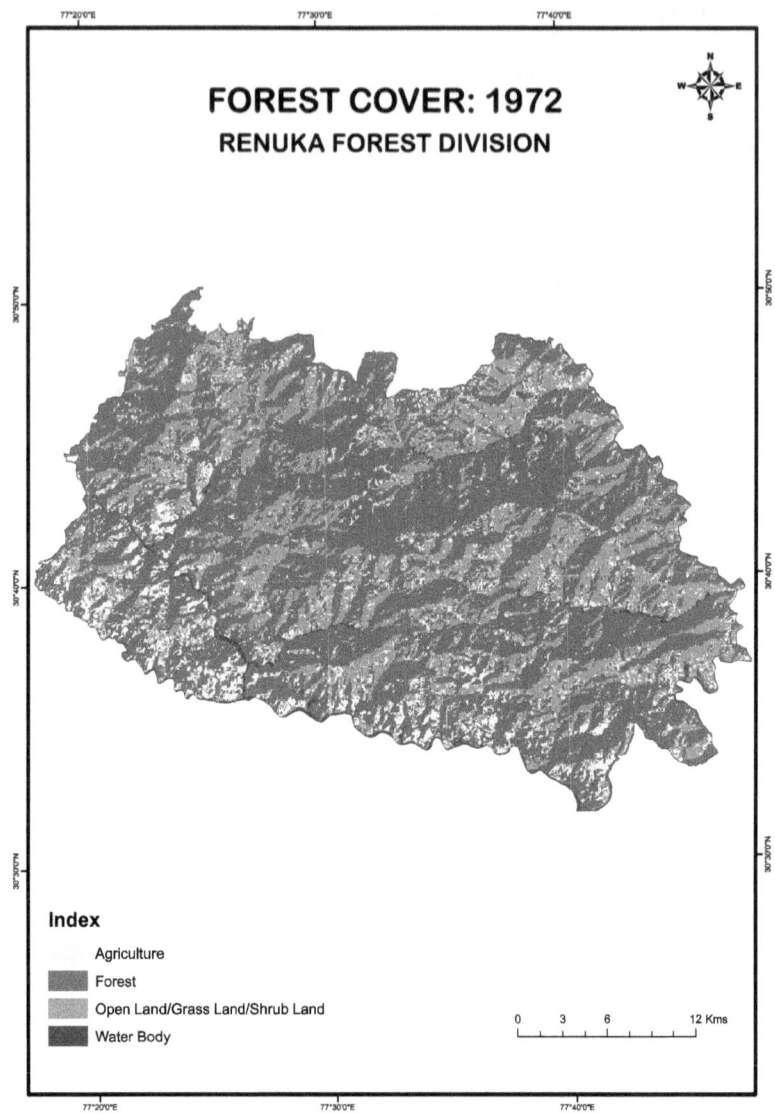

Fig. 4.1

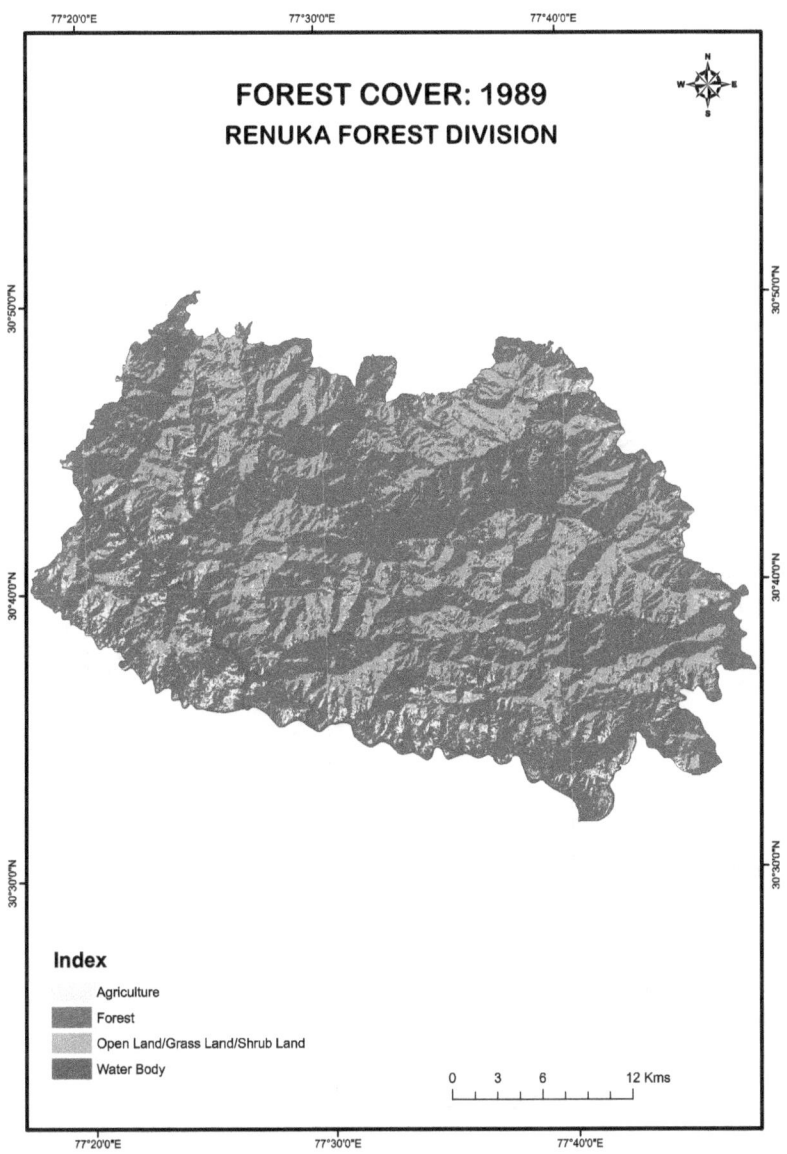

Fig. 4.2

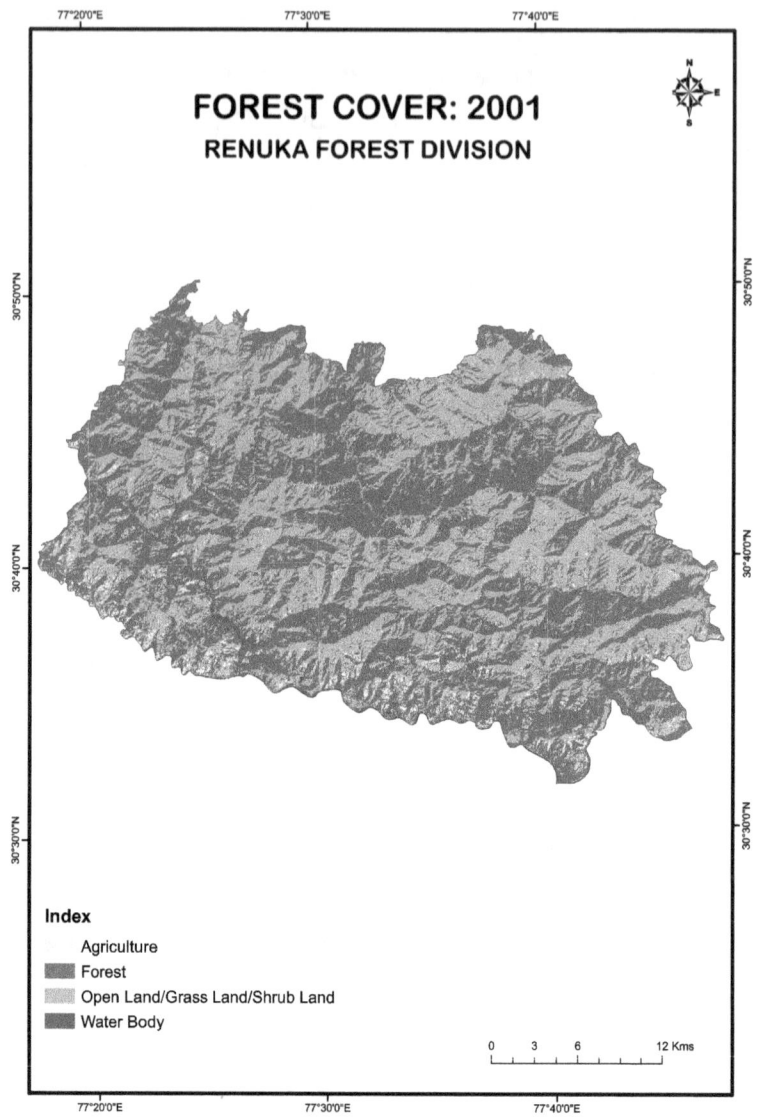

Fig. 4.3

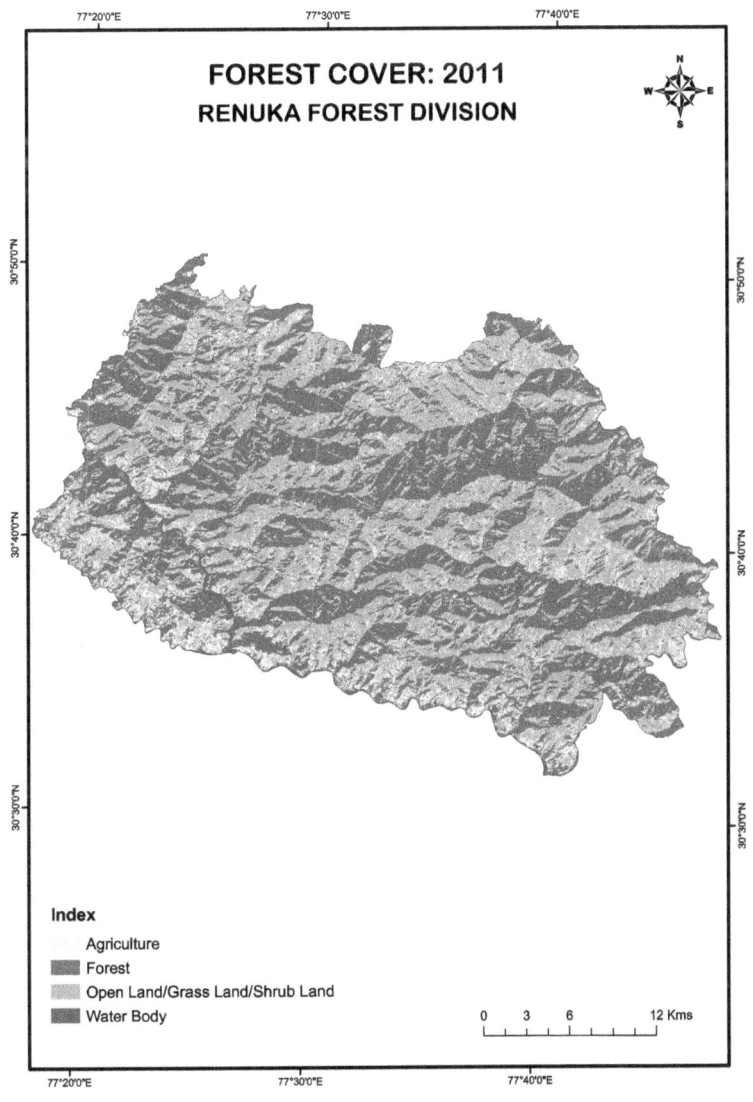

Fig. 4.4

4.4 Forest Cover Changes of the Study Area from 1972-2011

In the study area **(Fig. 4.5)**, forests, which were occupying 610 sq.km in 1972, is found to occupy 634 sq.km in 1989. The forests, which occupied cover in 61 percent for the study area in 1972 increased to 64 percent in 1989. All these observations clearly prove that during the period 1972-1989 forests have increased intensively due to efforts were underway to restore and rehabilitate degraded areas by bringing them under massive afforestation, social forestry and fuelwood/fodder development programs. It also increased due to afforestation programme carried out by forest department mainly in Renuka range (Ghataun), Kafota range (Tatiyana), Shillai Range (Shri Kyari, Chyali, Bhatnaul, Kota pab, Khatva, Milla, Jaswi, Lani, Baror, Dabar, Jarwa, Jakandon, Naipanjor, Tatwa Beyong), Sangrah Range (Daskana, Taikri, Panjah, Bhaltar, Lajwa, Arat, Ranphuwa, Uncha Tikkar) and Nohra Range (Manal, Chokar, Pipli, Bandal, Shilli, Bhangar, Bhangari, Nohra, Bhog, Charna, Ghandoori, Chunvi and Sail). In these forest ranges Deodar and Kail were planted. However, the natural growth in tree cover has been also noticed during the field survey. In the study area, open land/grass land/shrub land, which occupied 190 sq.km. during 1972 got increased to 230 sq.km in 1989. In terms of percentage, open land/grass land/shrub land, which occupied 19 percent of the study area in 1972, got increased to 23 percent in 1989. Thus it is clear that during the period 1972-1989, open land/grass land/shrub land have increased. This indicates that there was no human interference in the hill ecosystem during this period.

In the year 1989, the forests occupied 634 sq.km, which is about 64 percent of the study area and the open land/grass land/shrub land was 230 sq.km, which makes about 23 percent of the area **(Fig. 4.6)**. The forests have been decreased to 558 sqkm, 56 percent in 2001. This shows the massive loss of forests as about 76 sq.km area of forests declined during the period 1989-2001. During the period 1989-2001 forests have decreased due to illegal encroachments by villagers and the forest cover in certain region has depleted at a faster rate as a result of over exploitation for meeting the daily human needs of fuel fodder and fibre **(Photo plate 4.1a, b)**. The decline in the forest cover has been the result of lopping and chopping for fuel wood and other purposes and growth of roads and other infrastructural facilities.

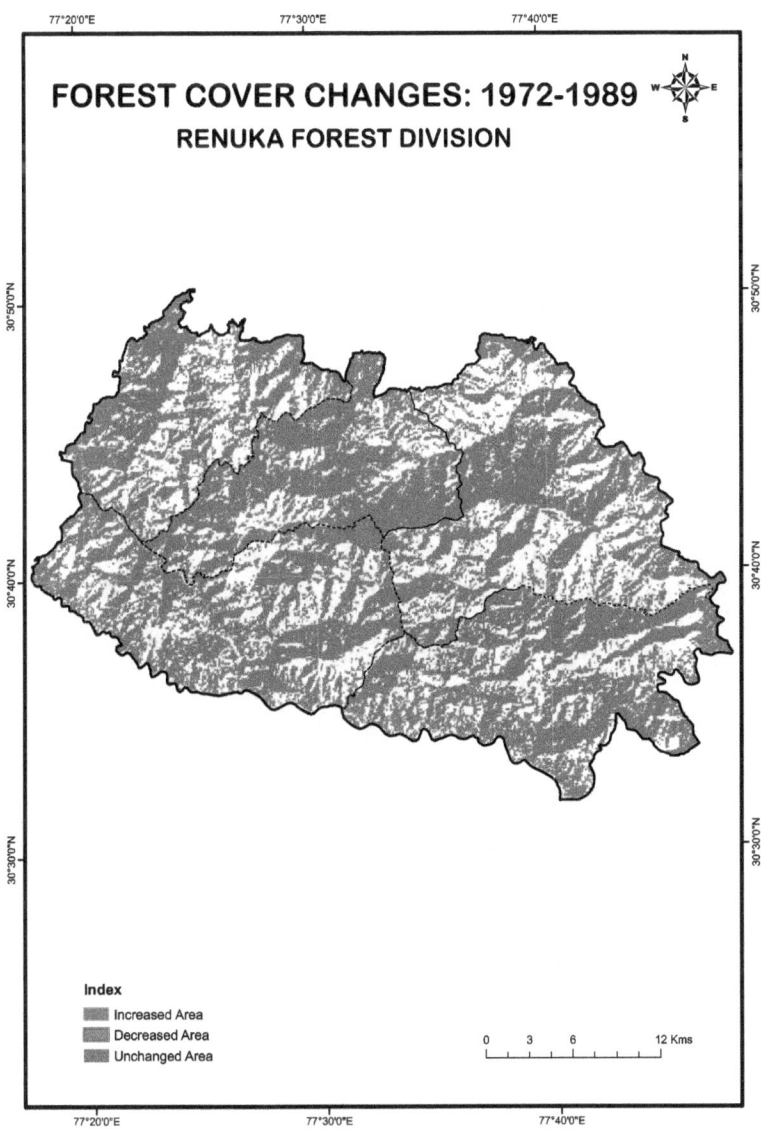

Fig. 4.5

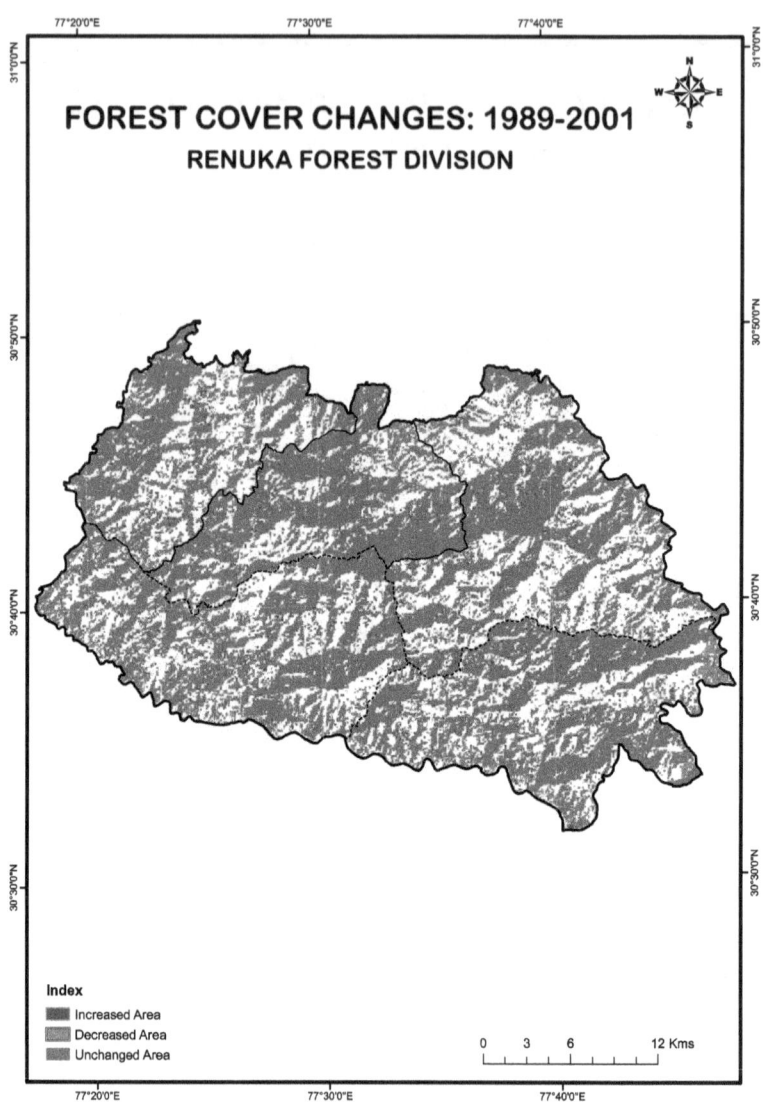

Fig. 4.6

However in 2001, open land/grass land/shrub land occupied 320 sq.km, covering about 32 percent of the study area about 90 sq.km area is occupied by open land/grass land/shrub land during the period 1989-2001. The analysis shows that during the period 1989-2001, open land/grass land/shrub lands have increased.

Photo plate 4.1a Piles of fuel wood for winter consumption **Photo plate 4.1b** Fodder used for Animals

The forests, which were occupying 558 sq.km in 2001, is found to occupy 549 sq.km in 2011(**Fig. 4.7**), registering a decline of about one percent to total forest cover. The study reveals that this decline in forest cover may be attributed to intensive agriculture activities emerging due to growing human pressure. The field survey revealed that some of the areas have witnessed large-scale depletion and degradation of forest cover. The mention may be made of Charag, Ganu, Cho Boghar forests in Renuka Range, Khajuri, Jamna Pabar forests in Kafota Range, Balokothi, Koti Bonch, Kharkhan, Loja, Manal, Bandauli forests in Shillai Range and Jamal Nihog, Bhajond forests in Nohra Range. The open land/grass land/shrub land, has registered a sharp decline from 320 sq.km (32 percent) to 242 sq.km (24 percent) during 2001 to 2011.

It is also inferred that forests cover in the study area has also been notably changed from 1972 to 2011 (**Fig. 4.8**). It also indicates that the area under agriculture and human habitation has substantially increased. Extensive damage to forests has been also caused by open grazing of cattle by local people (**Photo plate 4.2**) and nomadics, Gujjars coming from lower shiwalik of Sirmour district and shepherd coming from Kinnaur. These cattle not only damage the new saplings but also make the soil under their hoof compact and prevent new sprouting. Nomadic people practicing transhumance cause widespread damage to hill forests. The construction activities in the form of buildings, means of transport and communication, dams, installed hydropowers (**Photo plate 4.3**) and reservoirs and mining and quarrying (**Photo plates 4.4**) have adverse impact on the forest lands.

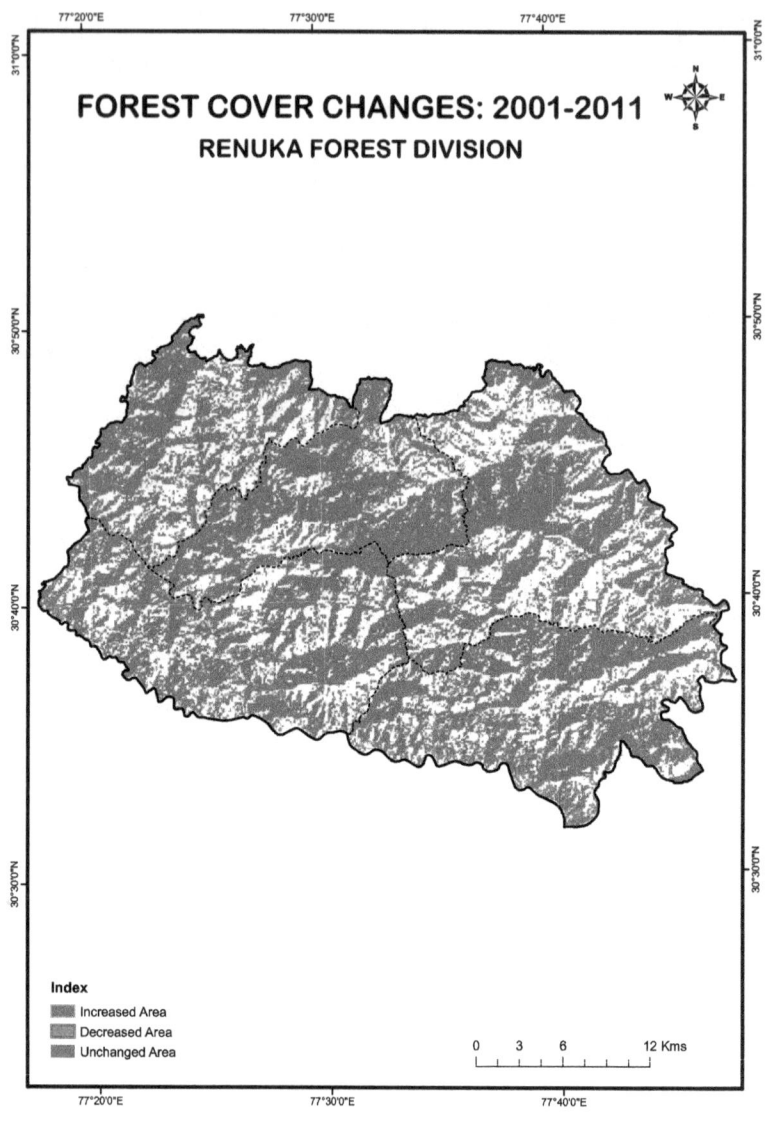

Fig. 4.7

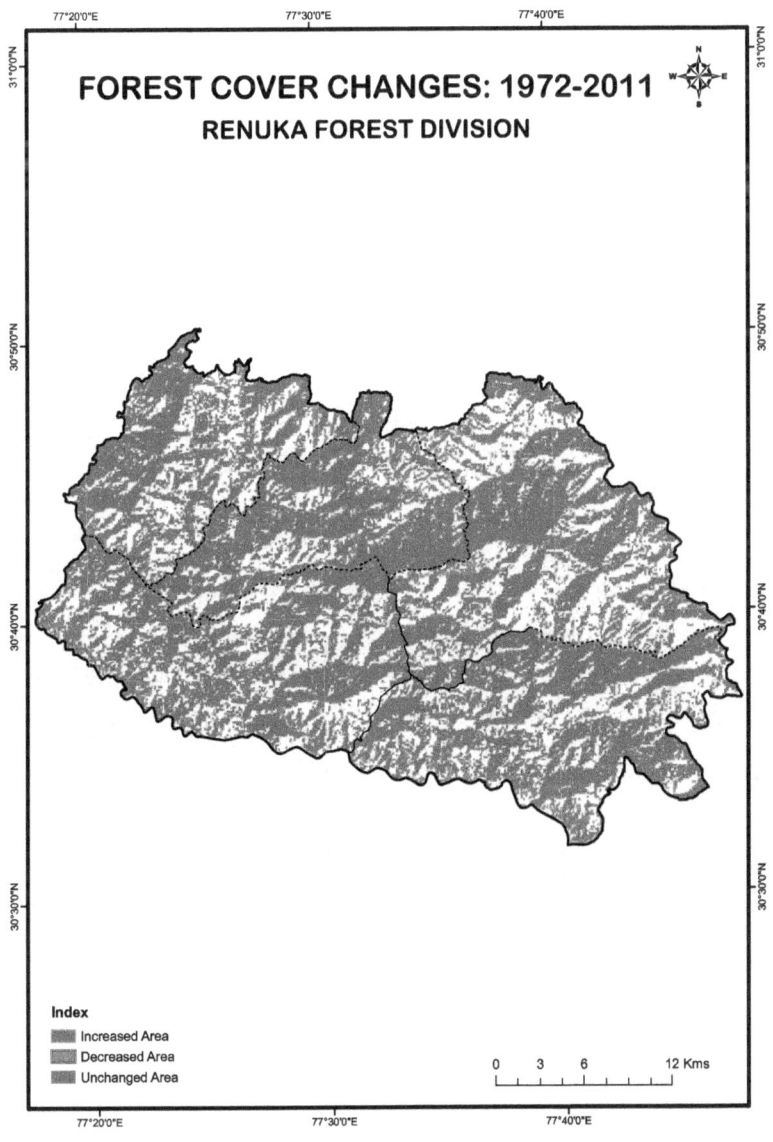

Fig. 4.8

Commercial activities like resin extraction, **(Photo plate 4.5)** oil extraction, fruit guarding and plantation also lead to massive deforestation. The massive encroachment has reduced the forest to few relict pockets. Due to excessive biotic pressure, heavy exploitation for the purpose of timber, fuelwood extraction, grazing and other local uses, the forest cover has been reduced and many areas are degraded. The forest cover in the region incurred major losses during last decade due to increase in population.

Photo plate 4.2 Grazing of Livestock

Photo plate 4.3 Installed Hydropower at Timbi

Photo plate 4.4 Mining Activities around Kamrao Village

Photo plate 4.5 Resin Oozing out of rills from Chil Forest

4.5 PEOPLE'S PERCEPTION ABOUT FOREST COVER CHANGES OF THE RENUKA FOREST DIVISION

Forest is crucial renewable natural resources and has an imperative role in preserving the environment suitable for human life. In India, forestry activities are closely related to the needs and survival of rural peoples. Dependency on fuel wood for cooking and house heating, fodder collection and grazing are traditionally practiced for livestock production, are main need of forest. **Fig. 4.9**, reveals about density of forest cover of the study area. More than half of people indicated that forest cover is dense while one fifth of the family stated that forest cover is open (Sparse) and remaining 10 percent of the respondents said about forest cover are bushes and 10 percent of families are of the opinion that there is no forest. This is mainly because these respondents belong to the areas like Kamrao, Kafota, Haripur Dhar, Sangrah, Ludhiana etc., which are devoid of any forest cover.

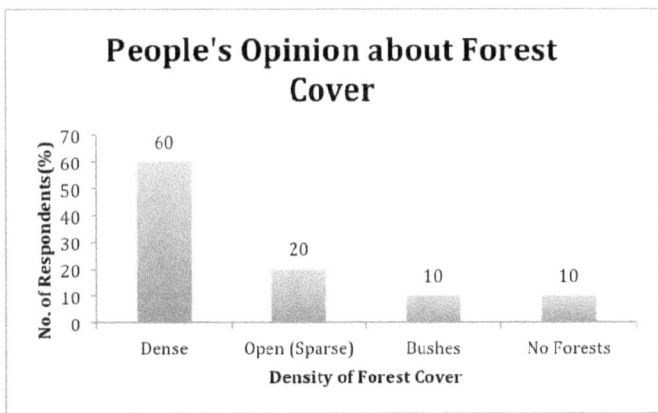

Source: Field Survey May, 2013

Fig. 4.9

Table 4.2, shows about change of forests in the past few years, half of the locals stated that forest has drastically decreased due to the growing need and greed of people over the past centuries. The uncontrolled felling of trees, excessive mining, construction of roads, clearing of vast area of land for agriculture, grazing large cattle population are some of the major reasons responsible for dwindling forest cover.

Table: 4.2 Change in Forest Cover during the Period of Four Decades

Forest Cover	No. of Respondents
Forest cover increased	20
Decreased	50
Not changed	10
Types of forests have changed	5
Biomass changed	5
Any other	10
Total	100

Source: Field Survey May, 2013

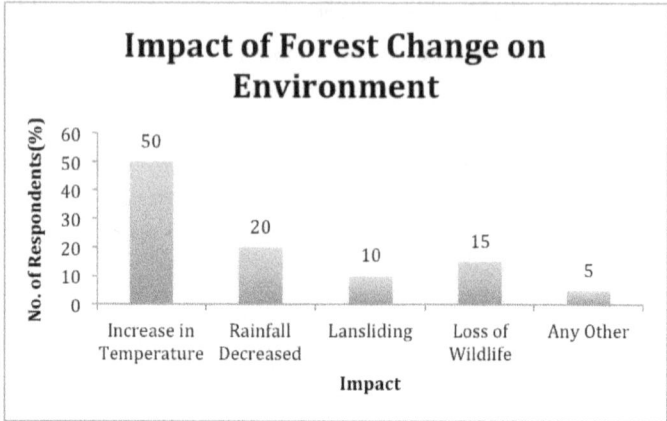

Source: Based on People's Perception, Field Survey May, 2013

Fig. 4.10

Fig. 4.10, depicts the impact of forest change on the environment. The half of the respondents opined that temperature has increased due to a change in the type, distribution and coverage of vegetation may occur given a change in the climate. Some changes in climate may result in increased precipitation and warmth, resulting in improved plant growth and the subsequent sequestration of airborne CO_2.

Table: 4.3 People's Opinion about Forest Resources in the Study Area

Opinion	No. of Respondents
Should be cut to earn money	20
Should not be disturbed	30
Should be raised more	40
Should be cleared for agricultural land	10
Total	100

Source: Field Survey May, 2013

A gradual increase in warmth in a region will lead to earlier flowering and fruiting times, driving a change in the timing of life cycles of dependent organisms (Sharma, 2005). In the opinion of one fifth of the respondents stated that rainfall has decreased whereas 10 percent of locals stated that land sliding has occurred due to change in forest cover and other factors i.e. mining, quarrying, deforestation, land use pattern, construction of roads and settlements and 15 percent of the respondents stated that loss of wildlife has occurred.

Table 4.3, shows that 40 percent respondents are of the opinion that forests should be raised because they supply oxygen to the atmosphere and therefore acts as lungs of the earth. They are rich in flora and fauna and form the home of rare species of plants and animals. Trees and forests have always been a life support system, not only for humans, but also for wild and domesticated plants and animals. They protect the land from erosion. The analysis of survey shows that 30 percent of families are of the view that forest should not be disturbed while one fifth of people stated that forests should be cut to earn money. The remaining 10 percent of locals have favoured the clearing of forest in order to make room for agriculture. They were uneducated and poor people of the study area that found during the field survey.

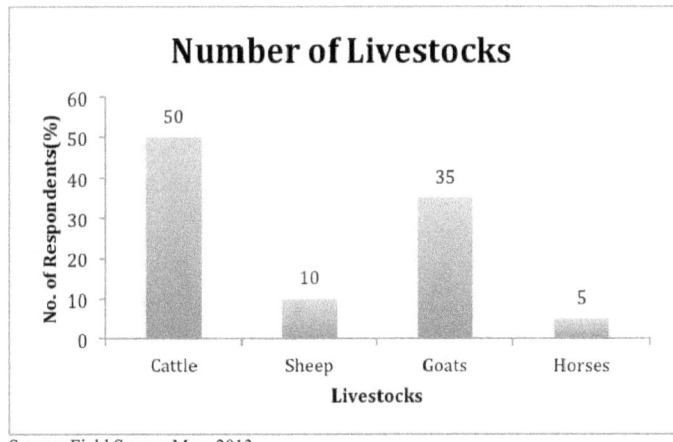

Source: Field Survey May, 2013

Fig. 4.11

Some of the households were seen rearing cattle as a side occupation after agriculture as an income source. Livestock's such as goat, hens, oxen, buffaloes, cow etc. are domesticated in their houses. Forests serve as the major source of fodder for these livestock. The local Zamindars are allowed to graze their buffalos, goats, sheep and other animals maintained for domestic purposes only on payment of grazing fees and all cattle kept for agricultural purpose free of cost in specified forest areas as a concession as per the Faisla-e-janglat (Working Plan of forest, 2014). **Fig. 4.11**, shows that number and types of livestocks, people have nearly half of cattles whereas 35 percent have goats and 10 percent of the respondents have sheep and 5 percent of families have horses.

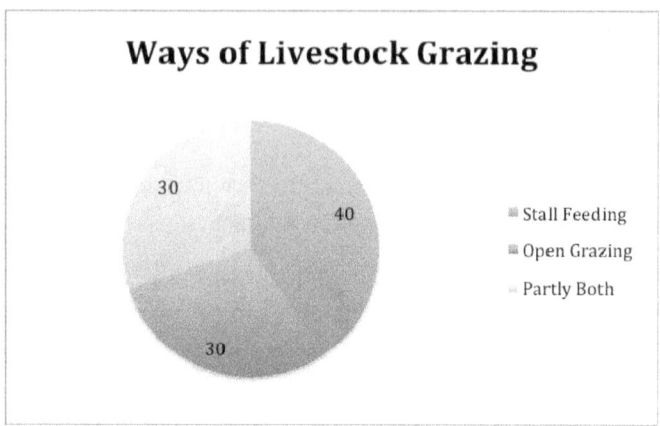

Source: Field Survey May, 2013

Fig. 4.12

There are two types of grazing prevalent in the study area, one of which is local grazing this is practiced by the villages living in and around the forest. This type of grazing is round the year. The cattle are left free in the forests in the morning and the cattle come back home in the evening. Another type grazing is nomadic grazing. This is of two types and can recognized as summer grazing and winter grazing. The summer grazing runs are mainly in the fir spruce areas. These are visited by Gujjars in the summer. Winter grazing is done by the shepherd coming from Kinnaur. **Fig. 4.12**, shows different ways of livestock grazing, less than half of people practices stall feeding while 30 percent of the respondents practices open grazing and they believed that it controls the growth of the non-native grasses and herbs so that other desirable plants (wildflowers and native grasses) can regenerate and coexist with them. Many plants, including several endangered species, require grazing to maintain viable populations. And remaining 30 percent respondents opines that they practices partly both. The issue of livestock as a part of major policy cannot be ignored especially when dealing with problems of deforestation, land degradation, climate change, and air pollution, water shortage and loss of biodiversity.

Table: 4.4 Major Threats to Green Cover of the Study Area

Threats	No. of Respondents
Forest fires	10
Livestock's grazing	20
Human beings	50
Forest smugglers	10
Mining activity	10
Total	100

Source: Field Survey May, 2013

Table 4.4, tries to accentuate the major threats to green cover. It shows that half of the respondents believe that human beings themselves are major cause of deterioration to the forest cover since the collection of fuel wood, fodder and other forest products may go beyond their regenerative capacity (Sharma, 2005). On the other hand 1/5 of the respondents feels that livestock's grazing increases soil erosion and deforestation. The rest of 30 percent responents have been trifurcated at the equal share of 10 percent opining that mining, forest fires and forest smugglers have been the major culprits and threats to green cover.

The forests are important natural resource that can potentially be sustainably harvested and managed to yield a diversity of commodities of economic importance. Wood is by far the most important product harvested from forests. The wood is commonly manufactured into paper, lumber, plywood, and other products. In addition, in most of the forested regions of the less-developed world firewood is the most important source of energy used for cooking and other purposes. Potentially, all of these forest products can be sustainably harvested. Unfortunately, in most cases forests have been overharvested, resulting in the mining of the forest resource and widespread ecological degradation and it is imperative that in the future all forest harvesting is conducted in a manner that is more responsible in terms of sustaining the resource.

CONCLUSION

The present study is an integrated approach of remote sensing, GIS and analysis of socio- economic data used for forest cover changes. This study has showed the utility of satellite images and GIS to monitor changes in the forest cover of the study area. The result shows that most of the forest cover has been under the human pressure depleting and degrading its originality over the years. The human impact on physical environment also discussed in the next chapter.

REFERENCES

Bawa, R. (1986): Structural and Functional Studies of Three Semi-Grassland Communities near Shimla. Ph.D. Thesis, H.P. University, Shimla, PP 57-58.

Beard, J.S. (1944): Climax Vegetation in Tropical America, *Ecology*, 25(2).

Bhaduri, B.M. et al. (2009): Spatio-Temporal Visualization for Environmental Decision Support in Geospatial Visual Analytics: Geographical Information Processing and Visual Analytics for Environmental Security (Eds: R. De Amicis), NATO Science for Peace and Security Series C: Environmental Security, Springer Netherlands, 5: PP 331-341.

Brown, W.H. and Curtis, J.T. (1952.): The Upland Conifer Hardwood Forests of Northern Wiscosin, *Ecological Monographs,* 22: PP 217-234.

Champion, H.G. (1936): A Preliminary Survey of the Forest Types of India and Burma, *Indian Forest Records,* (N.S) Silva.

Champion, H.G. and Seth, S.K. (1968): A Revised Survey of Forest Types of India, *The Manager of Publication,* New Delhi.

Chand, U. (2002): Floristic Composition of Some Parts of Spiti Valley, M.Phil Desertation, Dr Yashwant Singh Parmar University of Horiticulture and Forestry, Nauni, Solan H.P. PP 17.

Connell, J.H. and Cris, E. (1964): The Ecological Regulation of Species Diversity, *American Nature,* 48: PP 399-414.

Curtis and Machintosh, (1951): The Interrelations of Certain Analytical and Systematic Phytosociological Characters, *Ecology,* 31: PP 434-496.

Curtis, J.T. (1959): The Vegetation Wisconsin: An Ordination of Plant Communities, University Wisconsin Press, Madison, PP 657.

Curtis, J.T. and Cottom, G. (1956): Plant Ecology Work Book Laboratory Field Reference Manual, Burgess Publishing Company, Minnesota, PP 193.

Dansereau, P. (1960): Origin and Growth of Plant Communities, In: Growth In Living Systems, *Procedings of International Symposium* held at Purdue University, Basic Book Inc., New York (1961), PP 567-603.

Dengiz et al. (2009): Soil Erosion Assessment Using Geographical Information System (GIS) and Remote Sensing (RS) Study from Ankara-Guvenc Basin, Turkey *Journal of Environmental Biology,* 30: PP 339-344.

Dwivedi, A.P. (1993): Forests the Ecological Ramification, Natraj Publishers, Dehradun, PP 19-32.

Egbert, S.L., Park, S. and Price K.P. (2002), Using Conservation Reserve Program Maps Derived from Satellite Imagery to Characterize Landscape Structure, *Computers and Electronics in Agriculture,* 37, PP 141-56.

Ehrlich, P. and Ehrlich, A. (1981): Extinction: the Causes of the Disappearance of Species, Random House, New York.

Fizher, A.G. (1960): Latitudinal Variation in Organic Diversity, *Evolution,* 14: PP 64-81.

Fosberg, F.R. (1958): A Rational General Classification of Humid Tropical Vegetation, *Humid Tropics Vegetation,* PP 32-39.

Friedman, S.Z. (1978): Use of the Ratio Threshold Classifier in Mapping Urban/Non-Urban Land Cover from Landsat MSS Imagery Master's Thesis, University of Wisconsin, Medison.

Ghildiyal, S., Baduni, N.P., Khanduri, V.P. and Sharma, C.M. (1998): Community Structure and Composition of Oak Forests along Altitudinal Gradients in Garhwal Himalaya, *Indian Journal of Forestry,* 21(3): PP 242-247.

Gupta, R.D. and Banerjee, S.K. (1991): Problems and Management of Soils and Forest Resources of North West Himalaya, Mahajan Book Centre, Jammu, PP 1.

Haber, W. (1990): Using Landscape Ecology in Planning and Management, Changing Landscapes; an Ecological Perspectives (Ed.) I.S. Zonneveld; and R.T.T. Forman. Springer-Verlag, New York Inc., PP 215-232.

Joshi, N.K. and Tiwari, S.C. (1990): Phytosociological Analysis of Woody Vegetation along an Altitudinal Gradiation in Garhwal Himalaya, *Indian Journal of Forestry,* 13(4): PP 322-338.

Jessica, P.K. et al. (2001): Forest Change Detection in Kalarani Round, Vadodara, Gujarat a Remote Sensing and GIS Approach, *Journal of the Indian Society of Remote Sensing*, 29: PP 129-135.

Jung, M., Churkina, G. and Henkel, K. (2006): Exploiting Synergies of Global Land Cover Products for Carbon Cycle Modeling, *Remote Sensing of Environment*, 101: PP 534-53.

Kapoor, K.S. (2006): Biodiversity in Cold Deserts of Western Himalayas: Issues and Interventions, *Mountain Technology Agenda: Status Gaps And Possibilities*, A Book Published by Department of Science and Technology, New Delhi in Collaboration with HESCO, Dehradun, Bishen Singh, Mohinder Pal Singh, Dehradun, PP 522.

Kapoor, K.S., Verma, R.K., Vaneet, J. and Rawat, R.S. (2005): Performance of Different Tree Species in Limestone Mine Spoil, *Annals of Forestry*, 13(1): PP 79-83.

Kapoor, K.S., Subramani, S.P., and Rawat, R.S. (2005): Some Lesser-Known Tree Species of Himachal Pradesh: Distribution and Status, *Environment and Ecology*, 23 (2): PP 288-296.

Kapoor, K.S., Subramani, S.P. and Vaneet, J. (2005): Medicinal Plant Wealth in High Altitudes Including Cold Deserts of Western Himalayas: Their Taxonomy and Medicinal Plants and Health Care Trust, Jodhpur, PP 218.

Kershaw, A.K. (1973): Quantitative and Dynamic Plant Ecology, Edward Arnold Ltd., London, PP 308.

Knight, D.H. (1975): A Phyto-Sociological Analysis of Species Rich Tropical Forest on Barrow Colorado Island, Panama, *Ecological Monographs*, 45: PP 259-289.

Kumar, D. et al. (2012): Ethno-Medicinal Uses of Some Plants of Kanag Hill in Shimla, Himachal Pradesh, India, *International Journal of Research in Ayurveda and Pharmacy*, 3(2): PP 319-322.

Kumar, N. and Choyal, R (2013): Ethno-Medicinal Uses of Some Plants of Lower Foot Hills of Himachal Pradesh for the Treatment of Oral Health Problems and Other Mouth Disorders, *International Journal of Advanced Research*, 1(5): PP 1-7.

Kushwaha, S.P.S. (1985): Environmental Monitoring and Cyclone Impact Assessment on Sriharikota Island, India, Project Report, National Remote Sensing Agency, Hyderabad.

Kushwaha, S.P.S. (1990): Forest Type Mapping and Change Detection from Satellite Imagery, *ISPRS Journal of Photogrammetry and Remote Sensing*, 45: PP 175-181.

Kushwaha, S.P.S., Kuntz, S. and Oesten, G. (1994): Applications of Images Texture in Forest Classification, *International Journal of Remote Sensing*, 15(11): PP 2273-2284.

Lambin, E.F., Turner, B.L. and Helmut J. (2001): The Causes of Land Use and Land Cover Change: Moving Beyond the Myths, *Global Environmental Change*, 11: PP 261-269.

Lillesand, T.M. and Kiefer, R.W. (1987): Remote Sensing and Image Interpretation, John Wiley and Sons, New York.

Louisa, J.M.J. and Antonio, D.G. (2001): Parametric Land Cover and Land Use Classifications as Tools for Environmental Change Detection, *Agriculture, Ecosystems and Environment*, 91: PP 89-100.

Macarthur, R.H. (1965): Pattern of Species Diversity, *Biological Review*, 40: PP 510-533.

Macleod, R.D. and Congalton, R.G. (1998): A Quantitative Comparison of Change Detection Algorithms for Monitoring Eelgrass from Remotely Sensed Data, *Photogrammetric Engineering And Remote Sensing*, 64(3): PP 207-216.

Malia, W.A. (1980): Change Vector Analysis and Approach for Detecting Forest Changes with Landsat, Proc. Machine Processing of Remotely Sensed Data, LARS, West Lafayette, U.S.A.

Margalef, R. (1968): Perspectives in Ecological Theory, University Chicago Press. Chicago. PP 78-87.

Mishra, M.K. and Mishra, B.N. (1981): Association and Correlation of Plant Species in a Tropical Grass Land Communities, *Tropical Ecology*, 22: PP 88-89.

Mishra, R. (1968): Ecology Workbook, Oxford and IBH Pub. Co., Calcutta, PP 244.

Monk, C.D. (1967): Tree Species Diversity in the Eastern Deciduous Forests with Particular Reference to North Central Florida, *American Nature*, 101: PP 173-187.

Murali, K.S. et al. (2002): Joint Forest Management in India and its Ecological Impacts, *Environment Management Health*, 13: PP 512-528.

Natrajan, I. (1995): Domestic Fuel Survey with Special Reference to Kerosene, National Council of Applied Economic Research, New Delhi.

Negi, H.S. (2002): Studies on Woody Species Diversity of Sangla Valley (HP), M.Sc. Thesis. Dr. Y.S. Parmar University of Horticulture and Forestry, Nauni, Solan, PP 29-30.

NRSA, (1983a): Report on Environmental Studies through Remote Sensing and Mapping of Idukki Area in Kerala, *National Remote Sensing Agency,* Hyderabad.

NRSA, (1983b): Nationwide Mapping of Forest and Non-Forest Areas Using Landsat False Colour Composites for the Periods, 1972-75 and 1980-82, Report, *National Remote Sensing Agency,* Hyderabad.

Paterson, S.S. (1956): The Forest Area of the World and its Productivity, Gutesberg, Sweden.

Panigrahy, R.K. (2010): Forest Cover Change Detection of Western Ghats of Maharashtra Using Satellite Remote Sensing Based Visual Interpretation Technique, *Current Science,* PP 98.

Paul, S. (1981): Biomass Plantation-Energy Farming, *Journal of Scientific and Industrial Research,* PP 40.

Phillips, E.A. (1959): Methods of Vegetation Study, Henry Holt And Co. Inc., New York, PP 105-107.

Porwal, M.C. and Pant, D.N. (1989): Forest Cover Type and Land Use Mapping Using Landsat Thematic Mapper-A Case Study for Chakrata in Western Himalayas, Uttar Pradesh, *Journal of the Indian Society of Remote Sensing,* 17: PP 33-40.

Puri, G.S. (1960): Indian Forest Ecology, Oxford Book and Stationary Company.

Ralhan, P.K., Saxena, A.K. and Singh, J.S. (1982): Analysis of Forest Vegetation at and Around Nainital in Kumaun Himalaya, *Proceedings of Indian National Science Academy,* 84: PP 121-137.

Ravinder, N.H. and Hall, D.O. (1995): Biomass Energy and Environment-A Developing Country Perspective from India, Oxford University Press, Oxford.

Rawat, R.S. (2001): Phytosociological Studies of Woody Vegetation along an Altitudinal Gradient in a Mountain Forest of Garhwal Himalayas, *Indian Journal of Forestry,* 24(4): PP 419-426.

Risser, P.C. and Rice, E.L. (1971): Diversity in Tree Species in Oklahma Upland, *Forests Ecology and Management,* 52: PP 786-800.

Roy, P.S. (1991a): Tropical Forest Type Mapping and Monitoring, *International Journal of Remote Sensing,* 129: PP 2205-2225.

Roy, P.S. (1991b): Forest Cover and Land Use Mapping in Karbi Analog and North Cachar Hills Districts of Assam Using Landsat MSS Data, *Journal of the Indian Society of Remote Sensing,* 19: PP 113-123.

Roy, P.S and Giriraj, A. (2008), Land Use and Land Cover Analysis in Indian Context, *Journal of Applied Sciences,* 8(8): PP 1346-1335.

Royal, J.A. (1980): Change Detection Method Development: Census Urban Area Application Pilot Test, Final Report, General Electric Co., Beltsville.

Sabins, F.F. (1987): Remote Sensing Principles and Interpretations, W.H. Freeman And Co., San Francisco, USA, PP 696-697.

Saxena, A.K. (1979): Ecology of Vegetation Complex of North Western-Catchments of River Gola, Ph.D., Thesis, Kumaun University, Nainital, PP 484.

Saxena, A.K. and Singh, J.S. (1982): A Phyto-Sociological Analysis of Forest Communities of a Part of Kumaun Himalaya, *Vegetation,* 50: PP 3-22.

Shanbhar, G.Y. (1958): A New Method for Classification of the Climates of Arid and Semi- Arid Regions, *National Institute of Science (Biology),* 24 (3).

Sharma, D.D. (2005): Forests Economy and Environment, Kilaso Books, New Delhi.

Sharma, V.V.L.N. et al. (2001): Land Use and Land Cover Change Detection Through Remote Sensing and its Climatic Implications in the Godavari Delta Region, *Journal of the Indian Society of Remote Sensing,* 29: PP 86-91.

Sharma, C.M. and Kumar, A. (1992): Community Structure of Some Natural Forest Stand in Lansdowne Forest Range of Garhawal Himalaya, *Journal of Tropical Forest Science,* 5: PP 8-12.

Shepard, J.R. (1964): A Concept of Change Detection, *Photogrammetric Engineering and Remote Sensing,* 30: PP 649.

Simpson, E.H. (1949): Measurement of Diversity, Nature (London), PP 163-688.

Singh, A. (1989): Digital Change Detection Techniques Using Remotely Sensed Data, *International Journal of Remote Sensing,* 10(6): PP 989-1003.

Singh, J.S. and Singh, S.P. (1987): Forest Vegetation of the Himalaya, *Botanical Review,* 53(3): PP 81-192.

Srivastava, B.P. (1980): Energy Plantation: A New Approach, Second Forestry Conference, FRI Dehradun, PP 16-19.

Stow, D.A., Tinnery, L.R. and Estes, J.E. (1980): Deriving Land Use/ Land Cover Change Statistics from Landsat: A Study of Prime Agricultural Land, Proc. 14th International Symposium on Remote Sensing of Environment, Ann Arbor, Michigan, PP 1227-1237.

Subramani, S.P., Vaneet, J. and Rawat, R.S. (2006): Plant Diversity Conservation Needs, *The Botanica,* 55: PP 45-59.

Sukumar, R. (1991): Long Term Monitoring of Vegetation in a Tropical Deciduous Forest in Mudumalai, South India, *Current Science*, 62: PP 608-616.

Thornthwaite, C.W. (1933): The Climate of the Earth, *Geographical Review*, PP 23.

Thornthwaite, C.W. (1948): An Approach Towards a Rational Classification of Climate, *Geographical Review*, 38(1).

Todd, W.J. (1977): Urban And Regional Land Use Change Detect by Using Landsat Data, *Journal of Research*, U.S.G.S. 6: PP 529-534.

Turner, J.A. (1961): The Uses of Evaporation Data and Theory in Forest Management, *Proceedings of Hydrology Symposium*, National Research Council of Canada, PP 22-48.

Unni, N.V.M. et al. (1985): Evolution of Landsat and Airborne Multispectral Data and Aerial Photographs for Mapping Forest Features and Phenomenon in a Part of Godavari Basin, *International Journal of Remote Sensing*, 6: PP 419-431.

Verma, R.K. et al. (2006): Comparative Evaluation of Plant Diversity and Soil Properties in Protected Plantation and Unprotected Degraded area in Kunihar Forest Division of Himachal Pradesh, *Geobioss*, 33(2-3): PP 113-118.

Verma, R.K., Kapoor, K.S. and Kumar, S. (2005): Status of Plant Diversity around Renuka Lake Wild Life Sanctuary Himachal Pradesh, *Environment and Ecology*, 23 (1): PP 158-163.

Verma, R.K. et al. (2005): Analysis of Plant Diversity in Degraded and Plantation Forests in Kunihar Forest Division of Himachal Pradesh, *Indian Journal of Forestry*, 28(1): PP 11-16.

Vimal, O.P. and Tyagi, P.D. (1986): Fuel Wood from Waste Land, Yatan Publications, New Delhi, PP 3.

Warming, E. (1909): Ecology of Plants, an Introduction to the Study of Plant Communities, Oxford University Press, London, PP 422.

Web, L.J. (1959): A Physiognomic Classification of Australian Rain Forests, *Journal of Ecology*, 47: PP 3.

Whittaker, R.H. (1965): Dominance and Diversity in Land Plant Communities. Science, 147: PP 250-260.

Whittaker, R.H. (1972): Evolution and Measurement of Species Diversity, Taxon, 21: PP 213-251.

Whittaker, R.H. (1975): Communities and Ecosystems, 2nd Edition, Macmillan Publication Company, New York, PP 385.

Working Plan of Forest (2014): Working Plan for the Forests of Renuka Forest Division, H.P. Govt., Forest Department.

Xiao, X.M., Zhang, Q. and Braswell, B. (2004): Modeling Gross Primary Production of Temperate Deciduous Broadleaf Forest Using Satellite Images and Climate Data, *Remote Sensing Environment,* 91: PP 256-270.

CHAPTER - V
HUMAN IMPACT
ON PHYSICAL ENVIRONMENT

The foregoing chapter was about forest types, changes in forest cover and people's perception about forest cover changes. The present chapter is based on impact of man on the physical environment and people's perception about forest sector policies.

5.1 IMPACT OF HUMAN ON THE PHYSICAL ENVIRONMENT

The human impact on physical environment can be realised in multifarious ways depending on the space, time and socio-economic activities. The most common anthropogenic impacts of forest fire, unscientific mining, exploitation of non-timber product and livestock grazing of the study area have been analyzed.

5.1.1 Forest Fire

Forests are subjected to various kinds of injuries out of which forest fire is the most consequential doing incalculable harms to the forests. After deforestation, forest fire is the most important cause of worldwide colossal damage to extensive forest area destroying seed, seedlings and young trees. Forest fire causes loss of timber, damage to life and property, loss of recreational value and destruction to wildlife etc. Repeated fires lead to partial or even total loss of woodlands and ultimately to the vegetation cover and humus. In India about 3 percent of the forest areas are affected annually due to fire and on an average over 34000 ha forest areas are burnt by fire every year (Anon, 1991). Fire has also been a major influence on the development of many of the world's forests and their management. Some forest ecosystems have evolved in response to frequent fires from natural causes, but most others are susceptible to the effects of wildfire, and every year millions of hectares of the world's forests are consumed by fire, resulting in enormous losses to the economy in timber burned and real estate degraded, damage to environmental, recreational and amenity values, high costs of suppression and even loss of life. The vast majority of today's fires in forest and woodlands are caused by humans, mainly as the result of the use of fire as a land management tool e.g. for conversion of forests into agricultural lands, for maintaining grazing lands or for facilitating the extraction of non-wood forest products. Forests are also burnt to clear the land for mining, industrial development or resettlement. Forest fires can result from personal and ownership conflicts (FAO, 2001).

Although fire has been the primary agent of deforestation, as a natural process it has an important function in the health and maintenance of certain ecosystems. Thus the traditional view of fire as a destructive agent requiring immediate suppression has given way to the view that fire can and should be used to meet land management goals under specific ecological conditions (FAO, 2001). Finally, density and the arrangement of a fuel influences flammability. If the fuel is close together, the fire can spread quickly-unless the fuel is packed so tightly that it cuts off air circulation which will slow the blaze. Weather conditions include wind, temperature, humidity and rainfall. Wind is one of the most important factors because it can bring a fresh supply of oxygen and also push the blaze toward a new fuel source. Temperature is important because fuels ignite and burn faster at higher temperatures. When humidity levels are low, which means there is less water vapour in the air, fuels will be drier and will ignite faster (Anon, 1999).

5.1.2 Main Factors Responsible for Forest Fire

The major factors or the catalysts of forest fires can be broadly grouped under the following categories.

5.1.2.1 Intentional: Forest fires are mostly anthropogenic in nature and caused intentionally. These may occur due to the following reasons (Negi, 1986).

- Forest floor is often burnt by villagers to get a good growth of grass in the following season or for a good growth of mushrooms.
- Wild grass or undergrowth is burnt to search for animals.
- Firing by miscreants.
- Attempt to destroy stumps of illicit fallings

5.1.2.2 Unintentional: These fires are caused due to man's carelessness i.e. without intention to set fire. Such fires may be due to the following reasons:

- Un-extinguished campfires of trekkers, labourers, camp of roadside charcoal panniers etc.
- Spark of fire from railway engines
- Careless throwing of fire after honey collection.
- Un-extinguished bidis, cigarette butts, matchsticks etc. By grazers, travellers, picnickers or even forest labourers.

- Burning of agricultural fields adjacent to forested areas. Such fires is left unattended may spread to forest areas.

- Careless handling of acid by resin tapers.

- During controlled burning by the department, fire may spread to the forest due to negligence of the staff.

5.1.2.3 Natural: Natural causes of fires include:

- Fires caused by lighting.

- Fires caused by rolling stones.

- Fires may be caused by volcanic eruptions.

5.1.3 Fire Hazard Mapping

Remote sensing has a tremendous scope in forest fire mapping. The first application of forest fire date from 1960 when several aerial infrared scanners were tested for fire spot detection (Chuvieco and Congalton, 1989). After the launch of earth resources satellites several studies have been done for forest fire and burnt area assessment (Tanaka et al. 1983; Rome and Despain, 1989). In addition to forest fire mapping remote sensing data has been effectively used for fire hazard rating system. In many fire hazard rating critical factors were vegetation, slope, aspect and elevation. Deeming et al. (1978) have used landsat MSS image to obtain fuel oriented vegetation maps. The remote sensing can help in three important ways in forest fire mapping, monitoring and management:

1. Active fire mapping

2. Burnt area mapping

3. Fire prone area mapping i.e. fire risk zonation

In Renuka forest division, the fire is common in the Chil areas **(Fig 5.1)**. However, the main cause of fire has been man and is due to negligence. The other reason of fire is the careless annual burning of the pastures by the locals wherein the fire escapes into the forests. The details of the fire cases in the study area are given in **table 5.1.**

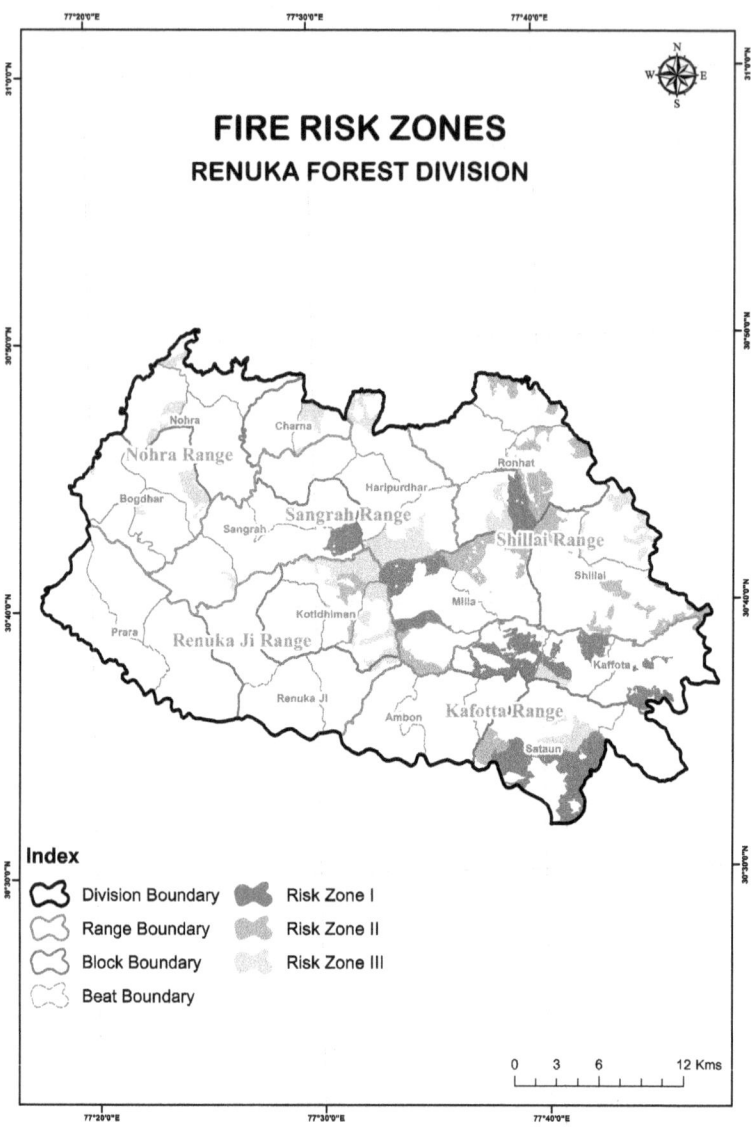

Fig. 5.1

Table: 5.1 No. of Fire Cases in the Renuka Forest Division (2009-2013)

Year	No. of fire cases	Area burnt (ha.)
2009-10	326	13939.56
2010-11	18	134.50
2011-12	6	63.00
2012-13	47	754.85

Source: Renuka Forest Division, 2013

5.1.4 People's Perception about Forest Fires and Management

Forest fire has been a source of disturbance for thousand of years. Forest and wild land fires have been taking place historically, shaping landscape structure, pattern and ultimately the species composition of ecosystems. The ecological role of fire is to influence several factors such as plant community development, soil nutrient availability and biological diversity. Forest and wild land fire are considered vital natural processes initiating natural exercises of vegetation succession. However uncontrolled and misuse of fire can cause tremendous adverse impacts on the environment and the human society. High terrain steepness along with high summer temperature supplemented with high wind velocity and the availability of high flammable material in the forest floor accounts for the major damage and wide wild spread of the forest fire.

The contribution of natural fires is insignificant in comparison to number of fires started by humans. The vast majorities of wild fires are intentional for timber harvesting, land conversion and socio-economic conflicts over question of property and land use rights etc. (Tanaka et al. 1983). The people's perception about forest fires has been depicted in **fig. 5.2**. It can be clearly understood that human negligence and some deliberate acts are the major reasons of forest fires. The negligence (60 percent) and deliberate efforts (30 percent) are both the results of human actions while just 10 percent are attributed to natural reasons. According to Forest Survey of India, about 6 percent of the forests are the prone to severe fire damage causing huge losses every year. Human beings often are responsible for forest fires that may be deliberate attempt or an accidental fire. Therefore, forest fire management, requires a balanced approach of a suitable mix of formal forest department input and local people's participation; with an appropriate use of technical and other locally adaptable social strategies for prevention and control of forest fires.

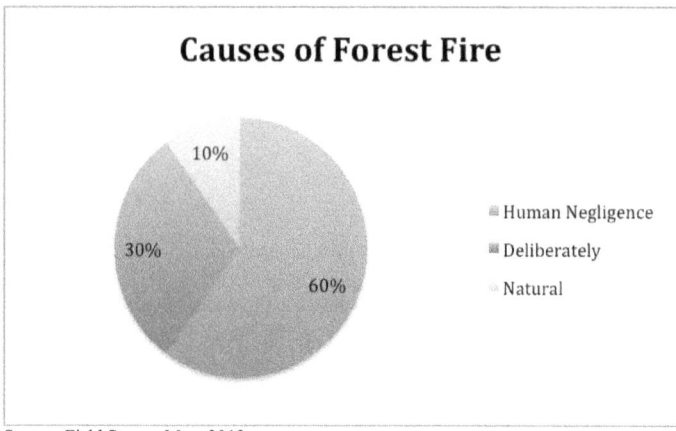

Source: Field Survey May, 2013

Fig 5.2

More than 19 percent of total forest area in Himachal Pradesh is highly prone to forest fires causing not only a huge financial loss to the State but also an equally huge loss to floral and faunal diversity. To control the very high incidences of forest fire, little less than half of people emphasize about the awareness of community whereas one fifth of the respondents demands proper legislation to curb the menace **(Table 5.2).**

Table: 5.2 People's Suggestion to Control Forest Fire

Opinion	No. of Respondents
Legal action	15
Proper legislation	20
Awareness of Community	45
Mixed views	20
Total	100

Source: Field Survey May, 2013

Another 15 percent believe that legal action against the fire setters may solve the problem and remaining 20 percent respondents holds varying and mixed views as a solution to control the forest fires in the study area.

5.1.5 Unscientific Mining or Quarrying

The mining of natural resources is invariably associated with land use and land cover changes (Prakash and Gupta, 1998). Therefore, mining is an important factor of anthropogenic influence on the environment, causing alteration of the landscape (Rigina, 2002). This also includes the land use and land cover change, urbanization and industrialization, land degradation and erosion. The extraction of minerals can cause increasing pressure on freshwater resources, agricultural land and forests. These resources are employed as inputs into economic development within the region, and the extraction of the resource itself is regarded as a stimulus for local and national development (Paull et al. 2006). Modern techniques of surface mining using heavy equipment can produce dramatic alterations in land cover, both ecologically and hydrologically (Simmons et al. 2008). Quantification of the effects that mining activities have on ecosystems is a major issue in sustainable development and resources management (Latifovic et al. 2005). Surface mining typically occurs in three stages, with different LULC conversions. In the first stage the site is cleared of vegetation and the uppermost soil horizons are removed and stored leading to homogenization of the material. The second stage consists of removal of soil and rock overburden, extraction of the mineral, and replacement of the homogenized overburden. The third stage is reclamation or re-establishment of vegetative cover. The homogenized soil is replaced, graded, and seeded. The conflict between mining activities and environmental protection has intensified recently, emphasizing the need for improved information on the dynamics of impacts at regional and local scales (Latifovic et al. 2005). Indeed, since the extent of surface mining in some areas has increased rapidly, new challenges for understanding the cumulative impacts to the physical and biotic landscape are generated (Shank, 2009). Knowledge of the extent of mining and reclamation within watersheds is for example critical to managing or mitigating the potential impacts of surface mining on downstream settlements (Townsend et al. 2009). Assessing cumulative environmental impacts is an important aspect of sustainable management and involves balancing benefits from resource exploitation against environmental degradation (Latifovic et al. 2005). Remotely sensed data are able to provide information on changes in local environments in a cost-effective way. In the case of environmental impact of mining, such data can potentially help in assessing changes in land cover, the extent of the physical impact of mining operations (infrastructure, mining pits, sedimentation etc.), as well as the effects of migration and settlement dynamics, particularly when used in conjunction with other forms of data (Paull et al. 2006).

The Renuka forest division consists of rich mineral wealth of limestone, barytes and lead. The main activity of mining in the study area is quarrying of limestone. The leases for mines of limestone were granted as early as 1963 and are still functioning (Working Plan of Forest, 2014). The demand for limestone is increasing and as such a constant efforts are being made to identify and bring additional areas under mining. Limestone mines of Renuka forest division are located in steep rugged and uneven mountainous terrain with steep slopes. Formations exposed in the area belong to lesser Himalayas, which are represented by high quality limestone, these formations are underlain and overlain by infra Krols and Tal formations respectively. These mines are located in Kamroo-Tillordhar, Bharli-Banor and Sangrah-Bhootmani belts of the study area. Total area under the limestone mining is 1975 acres as per the record of Renuka forest division **(Table 5.3)**. Working of the mines being practiced by open cast methods and with traditional tools and techniques in the study area.

Table: 5.3 No. of Mines, Status of Land and Areas in the Renuka Forest Division

Sr. No.	No. of Mines	Status of Land	Areas (Acres)
1	25	Private	503.11
2	19	Revenue	592.99
3	5	Forest	136.35
4	5	Partly Private and Revenue	216.81
5	1	Partly Private and Forests	526.01
		Total	1975.27

Source: Renuka Forest Division, 2013

Some mines have switched over to modern mechanical methods now. These mines are still using opencast method but dumpers/tippers, air compressors and drill machines like jackhammers along with combination of simple implements like crowbars chisels, pick axes etc. are also being used. Mining machinery like JCB are also practiced for excavation and roads construction. The mining activities causes the serious ecological problem to the study area that has been reflected in terms of degradation of vegetation cover, wild life, soil erosion, drying up of natural water springs etc. There have been no efforts to rehabilitate the areas being mined in the past. Neither is there legislation which makes the collaboration of using agency possible for reclaiming the areas, nor have any efforts been put to restore the vegetation in the mined areas.

5.1.6 People's Perception about Unscientific Mining or Quarrying

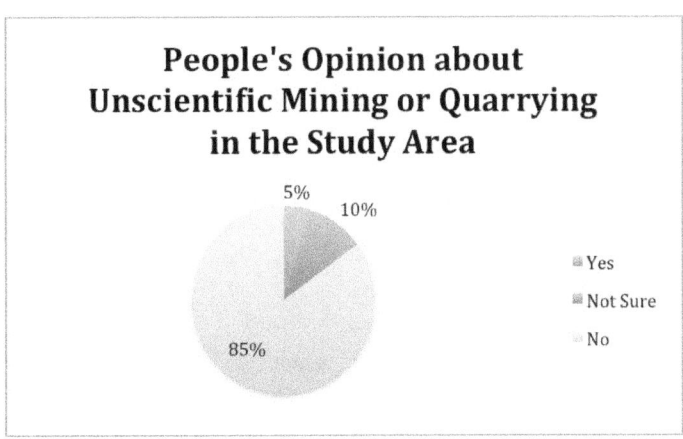

Source: Field Survey May, 2013

Fig 5.3

Diversion of forestland for mining purpose does results in loss of the forests in the area actually broken up for mining. However, to compensate the loss of forest land and mitigate the other adverse impacts, the diversion of forest land for mining purposes is subject to the creation and maintenance of compensatory afforestation, maintenance and regeneration of safety zone, soil conservation measures, phased reclamation of mined area to be utilized for regeneration and protection of forests. In the Renuka forest division of Sirmour district, reckless limestone mining has resulted not only in rapid destruction of forests, but also in water pollution. **Fig. 5.3** shows peoples opinion about unscientific mining or quarrying in forest area. A very high proportion of about 85 percent of people are not in favour to continue the mining in the study area. There have been just 5 percent people who support the continuation of mining. This clearly shows that mining activity has been benifitted a small section of the society while a very high majority is facing adverse impacts so they want to stop the mining activity in their surrounding.

5.2 EXPLOITATION OF NON-TIMBER PRODUCT

A forest is the store house of a large number of biotic components. Of all terrestrial system, the forest bio-mass represents the maximum richness and complexity in terms of numerous forest products. All forest products other than timber are called non-timber forest product or NTFP. These were earlier called minor forest products or MFPs. This was due to the fact that in the past timber was considered main object for the forest management. It was later realized that the so called minor forest products in fact play a major role in local as well as regional economy and the adjective 'minor' did not make much sense. Keeping in view the significance of non-wood forest products an attempt is being made in the present study to discuss, evaluate and understand the status of these forest products in Renuka Forest Division.

5.2.1 Medicinal Plants

Man has used medicinal plants as a major source of therapeutic agents for thousands of years. In India, out of 15,000 species of flowering plants about 17 percent are considered to be of medicinal value (Jain, 1968). Most of the plants are collected from nature indiscriminately for commercial exploitation, resulting in depletion and to some the extents are becoming rare and endangered. With the patronage of herbal medicines and their products increasing, there is an urgent need to conserve the endemic diversity in the medicinal plants before it is wiped out from nature. The factors contributing to the erosion of diversity are: bringing more land under cultivation due to biotic pressures, commercialization of agriculture for increased production, introduction of selected improved cultivators and thrust on socio-economic development. Therefore, considering these factors, an inventory of medicinal plant species which evolved and adopted over a long period, assumes great significance (Dalal et al. 1998; Kumar et al. 1997; Pushpangadan, 1998).

Information regarding the medicinal plants found in the study area has been gathered from practicing foresters as well as from the villagers who are engaged in collecting these medicinal plants. Analysis of these information is very useful in curing diseases, prospects and methods of practice and future guidelines for tapping of such medicinal plants (Mahato and Chaudhary, 2005). In the study area 16 plants species have been identified during the field survey and details are as below.

5.2.1.1 Abies spectabilis (Pinaceae): *Tosh* **(Photo plate 5.1)**

Part used **Leaves**

Warm decoction is prepared (about 20-30 fresh leaves boiled in a cup of water) and taken daily for 15 days to relieve chronic bronchitis (Chauhan, 1999).

Photo plate 5.1

5.2.1.2 Acacia catechu (Fabaceae): *Khair* **(Photo plate 5.2)**

Part used **Bark**

Warm decoction of matire wood (about 50 gm boiled in a cup of water) is taken 2-3 times a day for 3-4 days to relieve throat infection and cough paste of fresh root is applied on the joint once a day for seven days to treat rheumatism (Mahato and Chaudhary, 2005).

Photo plate 5.2

5.2.1.3 Aconitum deinorrhizum (Ranunculaceae): *Mohru*

Part used **Root**

The root is smoked in hooka to relieve acute gastric pain. Root powder mixed with mustard oil is used for massage for 3-4 months in paralytic body parts and rheumatic joints.

5.2.1.4 Bauhinia variegata (Caesalpiniaceae): *Kachnar*

Part used **Flower, Root**

Dried flowers, ground powdered mixed with mishri and butter is given to the patient for about 15 days in case of piles decoction of root is given twice a day for 3-4 days in case of snake-bite.

5.2.1.5 Berberis lycium (Berberidaceae): *Kashmal* (Photo plate 5.3)

Part used **Root**

Decoction of root with honey is given twice a day for 7-15 days in jaundice root extract is given twice a day for 3-6 month in case of leprosy.

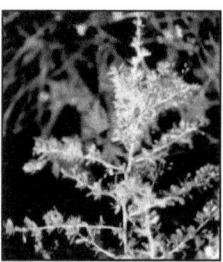

Photo plate 5.3

5.2.1.6 Berberis aristata (Berberidaceae): *Kushmol*

Part used **Bark, Root**

Rasount (Bark and root bark) mixed with honey is given to the patient twice a day for 5-7 days in urinary tract infection and piles.

5.2.1.7 Betula utilie (Betulaceae): *Bhojpatra* (Photo plate 5.4)

Part used **Papery bark**

The decoction of papery bark is prepared and is used as a vaginal douche to avoid conception.

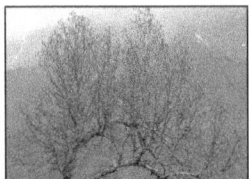

Photo plate 5.4

5.2.1.8 Butea monosperma (Papilianaceae): *Dhak Palash*

Part used **Leaves and Bark**

Fresh leaf juice is taken orally once a day for about 11-13 days in case of glycosuria. Dried powdered bark is taken twice a day for 7 days in case of menstrual disorder.

5.2.1.9 Cassia tora (Fabaceae): *Chokar*

Part used **Seeds**

Seeds with turmeric and mustard oil is made intopaste and applied on the affected area in case of eczema and skin diseases. One teaspoonful of crushed seed is taken with tea 2-3 times daily for a week to cure cough, headache and fever.

5.2.1.10 Centella asiatica (Apiaceae): *Brahmi* (Photo plate 5.5)

Part used **Leaves**

Powdered dried leaves are given to cure paramnesia and for improving memory. One or two leaves are taken every morning to overcome stuttering.

Photo plate 5.5

5.2.1.11 Cedrus deodara (Pinaceae): *Devdar* (Photo plate 5.6)

Part used **Wood**

The decoction of the bark is prepared and is given with turmeric and guggulu twice a day for 11 days in case of Gonorrhoea and Syphilis.

Photo plate 5.6

5.2.1.12 Dalbergia sissoo (Fabaceae): *Shisham* (Photo plate 5.7)

Part used **Leaves**

Decoction of leaves serves as a stimulant. It is also used as a blood purifier and to cure urinary tract diseases. Crushed leaves along with lime juice are used for once.

Photo plate 5.7

5.2.1.13 Duchesnea indica (Rosaceae): *Kaphal*

Part used **Leaves**

The leaves are crushed and applied as paste in skin diseases, wounds and cuts.

5.2.1.14 Fagopurum esculentum (Polygonaceae): *Ogla*

Part used **Root**

Decoction of the root is given twice a day for 10 days in case of typhoid.

5.2.1.15 Pinus roxburghii (Pinaceae): *Chil* (Photo plate 5.8)

Part used **Stem**

Oil extracted from stem is given twice a day for 5 days in case of constipation.

Photo plate 5.8

5.2.1.16 Rhododendron arboreum (Ericaceae): *Buras* (Photo plate 5.9)

Part used **Flowers**

Dried flower powder is given with water two times a day for 5 days in case of diarrhoea.

Photo plate 5.9

5.2.2 Resin Tapping

Resins are the complex oxidation products of various essential oils and vary in their chemical composition. They originate through reduction and polymerization of carbohydrates. Resin is insoluble in water but soluble in alcohol. They are inflammable and burn with smoky flame. It is exuded from the wounds of the bark well below the cambium in liquid state. However, on drying, resins generally remain liquid or in thick viscous state or some times it changes to solid form. Various studies have shown the use of adhesives in ancient times also (Nicholas, 1950). Resin is one of such natural adhesive finding application in different industries. It is used for manufacturing resin and turpentine oil, which in turn is of tremendous use in the industries like paper, incense, adhesives, printing inks, chewing gum, rubber, fragrance and flavour, cosmetics, disinfectants, soaps and detergents, camphor, ammunition, confectionary etc.

The state of Himachal Pradesh pioneered as a resin extractor at commercial level in the country. This is one of the important non-timber forest produce (NTFP) in the Renuka forest division. These forests were, infact, first to be tapped and used to be, in part, a major source of revenue of this division. Chir pine (Pinus roxburghii; family: conifers) is one of the six pines of India and is most widely distributed particularly in the outer Himalayas. It is the principal pine (*Pinus roxburghii*), which is being commercially tapped for resin in India. The tree is widely distributed in the outer ranges and principal valleys of the Himalayas, as also on the Shiwalik formations at elevation between 450 m to 2300 m. Tapping of resin is a sensitive problem as it involves the health of the tree. The tapping of resin needs to take number of precautions. While tapping the resin it should be seen whether tree being tapped is healthy and mature. Secondly the numbers of blazes are in line with the scientific principle laid down for the process. In the third place it is important to see that the excessive tapping is not being carried out. Last but not the least the method of tapping should be most modern and scientific. If all these precautions are taken into consideration, the ecological health of Chir pine forest will not be adversely affected otherwise these forests would become unhealthy, degraded, deformed and may suffer irreparable losses. There are different methods involved in the tapping of resin, which have been in use in our study area. Important among these are (a) French Cup and Lip method. (b) Rill method and (c) Bore hole method. All these method with their history of application in the study region are being discussed as under:

5.2.2.1 French Cup and Lip Method

In the Renuka forest division resin tapping was being done under "French cup and Lip" method. In this method a blaze has to be a size of 10 cm. wide and blaze is to be given to soft wood of the pine tree for extracting the resin. Depth of the blaze should be between 2 cm. to 2.5 cm. and length should be restricted to 48 cm. in the first year and in the subsequent year it should not be more then 38 cm. A lip is attached to the base of the channel, which directs the out flowing resin into the cup attached under the lip with the help of the nail experiment done in the Himachal Pradesh show that "the use of nail lips increased the yield by 10.9 percent on the northern aspects and 6.76 percent on the southern aspects over the conventional wood inserted lips (Raghav, 1971). This method is slow and continuous but does not affect the health of the tree if work is done according to the prescribed method.

5.2.2.2 Rill Method

FRI, Dehradun developed a method of resin tapping called "Rill Method" which was introduced in Himachal Pradesh in 1984 when 21,000 trees were tapped with this method in District Sirmour. Tapping by rill method is about 19 years old in this division. Because of the results achieved through and advantages of this method over French "Cup and Lip" method, more and more areas were brought under this method. It was in the year 1991 when the Himachal Pradesh forest department and H.P forest corporation completely switched to this new method of resin tapping

In this method a rill type blazes are made by freshening knife, which does not make more than 2 mm. deep blaze. There are series of such small blazes made on the trunk. The average width of the interspaces left between two consecutive rills is 5 mm and average width of the rill is 6 to 7 mm. There will be five blazes in a channel thus a channel can be tapped for five years. Interspaces between two channels will be 7.5 cms. And, at the base of the groove a lip is fixed with the help of a nail. A pot is attached under the lip or lying on the ground to collect the resin oozing out of these rills. The method envisages a tapping life of 20 years for the trees of 30 cm. diameter and above. Experimental evidences suggest that yield of resin in this method is slightly more than that from 'cup and lip' method. When the research was being conducted in Dehradun, the rill method was tried on a tree of 30 cm. diameter from March to October. The resin obtained per tree per blaze was 3.2 kg., where as with French Cup and Lip method it was 2.5 kg. per tree per blaze. Thus the rill method is not only production efficient but also suitable for maintaining the health of the tree and ultimately ecology of the Chir pine forest.

5.2.2.3 Bore Hole Method of Resin Tapping

The resin extraction techniques are consistently being improved to get maximum yield of resin with minimum loss of timber. One of these improved techniques of is the borehole technique. The extraction of resin by borehole technique has been described in detail by Hodges (1995). Holes are drilled into the tree's sapwood to open the resin ducts and collect the exuding resin in a closed container. Prolonged resin flow from boreholes for a period of several months is a key feature of this system, and represents a logical extension of the evolutionary trend in gum resin production systems toward internalization of the wounding process. Because borehole wounds cause relatively little damage to the tree's bark and cambium tissue, it is expected that there is less disruption of its normal growth processes. The holes are typically covered by new cambium and bark growth within about two years, and the tree may be allowed to continue growing without distortion of its form. Since holes are made near ground level, there is little or no damage to the merchantable part of the tree. The techniques could prove to be very effective in conservation and management of Pine resources in India.

Recently, "Bore Hole" method of resin tapping has been introduced in Himachal Pradesh. The resin extracted by Bore hole method is free from impurities and 'Sakki'. (At the time of extraction of resin, the bark chips, pine needles get mixed with the resin. This material in colloquial language is known as 'Sakki'). There is no loss of turpentine oil by evaporation as the resin is collected in polythene bags which are sealed. When this resin is processed the yields of pale grades of resin are much higher. The recovery is also high. Therefore this method when adopted on large scale will reduce losses and wastage to a greater extent. This method was tried on experimental scale for tapping 15,000 and 20,000 bores in Himachal Pradesh during 2000 and 2001 resin season. The maximum and minimum yield per bore hole on an average is 1.8 kg and 0.200 kg respectively. It can thus be said that the yield is not satisfactory. The technique has to be improved so that it becomes economical.

5.2.2.4 Production and Commercial Extraction of Resin

Resin extraction has been an important economic activity in Renuka forest division. Although this tapping was being done elsewhere in the Himachal Pradesh even before that. The production or extraction of resin from the Chir pine depends among other factor, on the aspects, on which a particular forest is located and the temperature that are being experienced at that aspect. Higher the temperature more would be the production and vise versa. In the study area temperature starts increasing from March onwards and allows the resin to be extracted till the end of October when temperature starts falling in the month of November the extraction is stopped.

Table: 5.4 No. of Blazes and Resin Extracted of Renuka Forest Division (1997-2007)

Year	Number of Blazes	Total Yield Obtained (Qtls.)
1997	22817	7192.86
1998	17848	5577.50
1999	15229	4759.06
2000	8600	2710.72
2001	6798	2124.37
2002	24116	7536.25
2003	33734	1070.10
2004	35191	1294.39
2005	44302	1894.09
2006	32115	1547.91
2007	39083	890.93

Source: Renuka Forest Division, 2013

Contribution of Renuka forest division has remained significant in the overall production of resin in Himachal Pradesh. The Shillai, Renuka and Kafota ranges contribute to the resin blazes. The extrraction is being done by the corporation on contract basis. The contractors import the labour from UP, Kangra, Hamirpur and Una because they are more skilled and the local labourers are generally not interested in resin tapping. The details of number of blazes and resin extracted year wise are as follows **(Table 5.4)**. The table reveals that the total yield has gone down drastically while the number of blazes has increased during the period 1997-2007. The tapping season varies from six to seven months depending on the weather condition. The on set of winter and prolonged monsoon decrease the tapping season. In lean season the yield is 31 quintals per section of 1000 blazes whereas in a good season the yield is 38 quintals per section of 1000 blazes.

5.3 LIVESTOCK GRAZING

Human and Natural environments have, presently, attained a problematic as well as interdependent relationship. Any further expansion of human environment through increase in number, results in catastrophic encroachment on natural environment, with unidirectional disturbing effects. Thus, the increasing pressure of over one billion human and half a billion livestock in India on the natural resources have constrained their productivity and reduced many areas as degraded. The extent of degraded land in India was 187.8 million hectare which is more than half of the geographical area of the country. As a result of all this an enormous scarcity of firewood (77 percent), timber (55 percent), green fodder (77 percent) and dry fodder to the tune of 51 percent has been created (Roy, 1999). The situation in Himachal Pradesh is no different and therefore efforts have been made to study the utilization of pastures in the Renuka forest division only. The animals rearing in the Renuka forest division vary from buffalo to sheep and goats. As such livestock plays a vital role in rural life. Every farmer thus, is forced to adopt an integrated livestock rearing and cultivation system. Feeding the huge number of livestock is not possible only through the fodder generated by agriculture, thereby making the grazing activity a crucial necessity. Most of the holdings in the study area do not permit the farmer to engage in fodder crop cultivation because of uneconomic returns and lesser duration of agricultural activities, due to unfavorable climatic conditions, uneven topography and steep slopes. Thus grazing happens to be an important economic activity of the region. For centuries, the alpine meadows in the region have been used as grazing grounds by migratory livestock of nomads as well as animals from adjacent lower valleys during summer months. According to the State Forest Department, the tree crops in the undemarcated, protected and unclassed forests is generally very open and large areas are absolutely denuded and used mainly as grazing grounds. No scientific evidence has been available on the past and present status of pasturelands in terms of their carrying capacities. But one can, after making a cursory look at these pastures, authoritatively say that these pasture and grasslands are being put to very heavy grazing. But there must be a limit to direct grazing, since too rapid a removal will kill the producers or greatly reduce their future productive capacity (Odum, 1971).

Further it is generally agreed by everyone that over grazing by livestock is identified as a prime factor leading to degradation of Himalayan ecosystem and in western Himalayas, the grazing pressure is estimated at thrice the carrying capacity of the grasslands and the system is under much stress at all elevations (Gupta, 1986).

Forest degradation in the Renuka forest division is largely due to ever increasing biotic pressure of man and his animals. Extent of grazing lands and pasture have been declining rapidly, the population of livestock has been increasing in leaps and bounds. The forest department has identified Sangrah, Shillai and Renuka ranges as ecologically the most fragile zones. The fragility to these areas has been brought about by animal gazing much in excess of the carrying capacity of the forests. The carrying capacity of grazing lands, in addition to the condition of soil; climate and vegetation type is basically determined by the inherent ability of vegetation tolerance under grazing capacity (Dasman, 1976). Village wastelands, which absorbed the pressure of grazing, have been converted into various uses other then grazing lands. The livestock, which earlier depended on the village pastures, are now directed to grazing in natural forests. As such grazing in forests is practiced all over in an unrestricted manner except for the closed areas. Alpine pastures and *Dhars* (A ridge) are excessively grazed by huge flocks of sheep's and goats of the local residents as well as the migratory grazers, resulting in deterioration of some of these areas (Working Plan of Forest, 2014).

5.3.1 Grazing system

In a hilly tract, as the study area is, particularly at higher altitudes, the soil is poor and stony and the slopes are too steep to render economic farming possible. The natural consequence is the population which cannot live on crop culture alone and they have taken to keeping of large herds of cattle and flocks of sheep and goats as allied activity. Unfortunately, due to wrong concepts, a trend has developed to carry out unrestricted grazing with an ever increasing number of cattle resulting from the division of families on the assumption that forests are the gift of nature and those living near these forests have the absolute right to avail this gift by use or abuse. However the graziers have been put into two categories in the study area viz. local and migratory.

5.3.2 Local Graziers

They live in villages and hamlets all over the division and stay there throughout the year. They are almost entirely agriculturists and keep buffaloes, cows, bullocks, mules, sheep and goats etc., and graze them in unclassed forests, demarcated, undemarcated, protected, and even in reserved forests where they have rights. A right holder can graze any number of cattle without paying any fee. This flow has resulted in heavy increase in the number of cattle as is borne out by the increasing cattle population. This has also resulted in tremendous increase in the number of uneconomical and scrub cattle.

5.3.3 Migratory Graziers

These graziers belong to the higher hills of the division. They stay in their houses for a limited period only and migrate to other places with their flocks to seasonal pastures and for the greater part of the year allow the animals to graze in area outside their barren lands. These graziers have got their agricultural holdings but these holding are insufficient to support the entire families and therefore, they keep large flocks of migratory sheep and goats which are principal source of their income. The migratory shepherds in the division are commonly referred to as Gujjars. The Gujjars who generally belongs to lower riches of Sirmour district. The rich pastures on the southern slopes of Shiwaliks provide autumn and summer grazing, but during winters the flocks move down to the lower hills areas, which provide only insufficient and poor grazing.

5.3.4 Livestock Population

The region is rich in livestock population. The livestock resources are varied in type and category and include cattle, buffaloes, goats and sheep, horses, donkey and other animals. As per animal census of 2002 the Division has a total 221896 livestock population, out of which 45 percent are cattle, 15 percent buffalo, 37 percent sheep and goats and 4 percent are other animals etc. In order to appreciate the magnitude of the problem of grazing vis-a-vis available resources and its impact on forest resources, it is essential to have an insight of the livestock population of the division.

5.4 PEOPLE'S PERCEPTION ABOUT FOREST SECTOR POLICIES

Forest rights in Himachal Pradesh are quite different from other parts of India. The colonial forest settlements in most parts of this state have recognized and recorded many local forest rights for several historical, economic, and political reasons. Village landowners have extensive user rights to graze cattle and collect fuel-wood, poles, and most non-timber products for their personal use. Most villagers also have the right to periodically harvest timber for house construction and repair. Additionally, many villagers can sell non-timber forest products and thereby benefit financially from what are today state forests. Thus, although almost all the forests belong to the state in terms of owner- ship, villagers enjoy extensive user rights to forests near their villages. Anderson's (1886) forest settlement of Kullu is a typical example. After much debate, the bulk of Kullu forests were classified under Chapter IV (protected forests) of the Indian Forest Act (1878), allowing considerable leniency in local people's use of forests.

Table: 5.5 People's Opinion about Legal Rules Pertaining to Forest Management and Use

Legal Rules	No. of Respondents
Fully Know	20
Partially Know	40
Little Idea	20
Not Sure	15
Do not Know at all	5
Total	100

Source: Field Survey May, 2013

All rights described above were registered in this forest settlement. Forest settlements in Himachal Pradesh are therefore progressive in the limited sense that they overtly recognize and legalize local forest uses. **Table 5.5** depicts that legal rules about forest management and use. It may be viewed that little above half of people are fully aware or knows partially about the forest management. About 20 percent respondents have little idea while 15 percent of people were not sure and remaining 5 percent do not know anything about forest management.

Table: 5.6 People's Opinion about Timber Distribution (T.D.) Rights

Opinion	No. of Respondents
Should continue as it is	50
Should be more liberal	30
Restricted	10
Should be banned	10
Total	100

Source: Field Survey May, 2013

Opinion about T.D. rights have been tabulated in **table 5.6**. It shows that half of the respondents were in favour and wants that T.D. should continue as per the present practice. They support their argument by saying that the study area is under high-risk seismic zone and the houses with wooden structures are best suited to meet such an eventuality. The hilly terrain has low road density thereby it is very difficult for the people to carry industrial material like cement, sand etc. meant for constructing such houses. The study area falls under the temperate zone with extremely cold weather conditions. In such an environment houses made with traditional wooden structures are more suited to live and can sustain for more than a hundred years. The another chunk of 30 percent of repondents are of the opinion that T.D. should be more liberal while 20 percent of people stated that T.D. should either be restricted and should be banned respectively.

Forests constitute 21 percent of the geographic area of India (SFR, 2009) and forestry represents the second major land use in the country after agriculture and it has been estimated that nearly 41 percent of the country's forest cover has been degraded to some degree (MoEF 2002). More than 14 percent of the population in India lives in the vicinity of forests (MoEF 2002), which provide both tangible and intangible benefits. The National forest policy 1894, was the first formal forest policy statement of India, and was based on the Voelcker Report in 1893 on Improvement of Indian Agriculture. The main stated objective of this policy was to manage the state forests for public benefit.

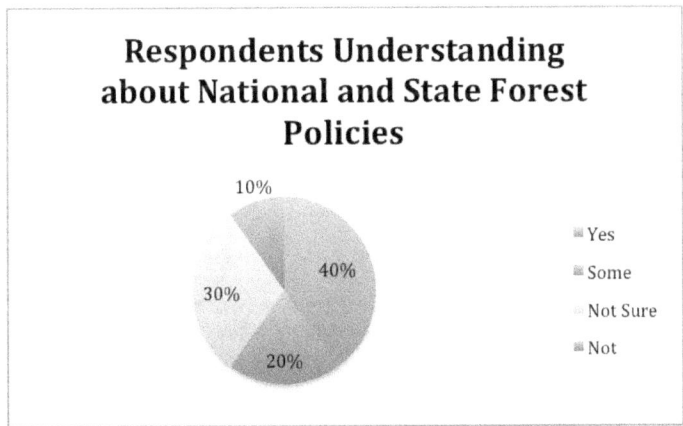

Source: Field Survey May, 2013

Fig. 5.4

The National forest policy of India in 1952 laid stress for the first time on having at least 33 percent of the National land area under forest cover. The current National forest policy of 1988 reiterated that the National goal should be to have a minimum of one-third of the total land area of the country under forest cover. In the fifth five-year plan (1974-79), the Planning Commission set a monitorable target of achieving 25 percent forest and tree cover by the end of 2007 and 33 percent cover by the end of 2012. According to National forest policy, 1988, at least two third i.e. 66 percent of the geographical area should be under forest in the hilly states like Himachal Pradesh. However, keeping in view that about one fifth of the area is inaccessible and beyond the tree limit, the State Government aims to bring half of the geographical area under forest cover. People's opinion about national and state forest policies has been shown through **figure 5.4**. It shows that about forty percent of people are aware about forest policies and another one fifth of people have some idea about these but 30 percent people do not carry any idea about the national or state forest policies. There have been about 10 percent respondents who holds same views about the policies but were not sure of their view point.

The Forest Development Agencies (FDA) has been conceived and established as a federation of Joint Forest Management Committees (JFMCs) at the Forest Division level to undertake holistic development in the forestry sector with people's participation.

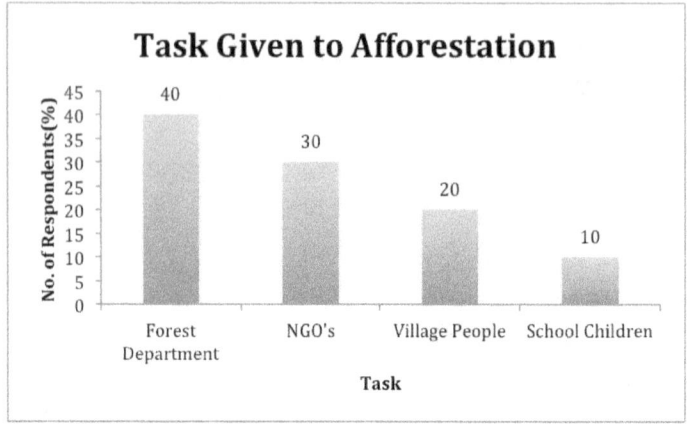

Source: Field Survey May, 2013

Fig. 5.5

This is a paradigm shift from the earlier afforestation programmes wherein funds were routed through the State Governments. This decentralized two-tier institutional structure (FDA and JFMCs) allows greater participation of the community, both in planning and implementation, to improve forests and livelihoods of the people living in and around forest areas. People's participation in any forest conservation is of vital importance. The **figure 5.5** depict about task given to afforestation. Less than half of people said that task should be given to forest department while 30 percent of locals believe in NGO's and one fourth of people stated that task may be given to village people and school children respectively.

CONCLUSION

The present study is mixed approach to analyze the interaction of human with physical environment and examine socio-economic data. The study has demonstrated the impact of forest fires, unscientific mining, resin tapping and livestock grazing in the study area. The underlying impact of human on physical environment shows that large-scale degradation and depletion of forests cover area has occurred in the study area. The land use/ land cover pattern shall be discussed in the next chapter.

REFERENCES

Anderson, A. (1886): Report on the Demarcation and Settlement of Kullu Forests (Reprinted 1975), Himachal Pradesh Forest Department, Shimla.

Anon, (1991): Indian Timbers, FRI Publications, Forest Research Institute (FRI and Colleges), Dehradun, India.

Anon, (1999): Aviation and Forest Fire Management, Forest Fire Fundamentals, Ontario Ministry of Natural Resources.

Anon, (1993): Forests of Himachal Pradesh, Department of Forest Farming and Conservation, Shimla, Himachal Pradesh, PP 4-7.

Badola, H.K. (2001): Medicinal Plant Diversity of Himachal Pradesh, Himalayan Medicinal Plants: Potential and Prospects, Himavikas Occasional Pusl No. 14, Gyanodaya Prakashan, Nainital, PP 407-430.

Badola, H.K. (2002): Endangered Medicinal Plants Species-Priorities and Action, Theme Paper, International Workshop on Endangered Medicinal Plant Species in Himachal Pradesh, G.B. Pant Institute of Himalayan Environment and Development, Mohal-Kullu, India (18-19 March), PP 11.

Brown, A.A. and Davis, K.P. (1973): Forest Fire: Control and Use, Mcgraw Hill, New York.

Central Council for Research in Ayurveda and Siddha (1990): Phytochemical Investigation of Certain Medicinal Plants Used in Ayurveda, *CCRAS*, GoI, New Delhi.

Champion, H.G. and Seth, S.K. (1968): A Revised Survey of Forest Types of India, Manager of Publications, Govt. of India, New Delhi.

Chaturvedi, A.N. (1981): Big Forest Loss Every Minute the Daily National Herald, Lucknow, PP 22-38.

Chauhan, N.S. (1999): Medicinal and Aromatic Plants of Himanchal Pradesh, Indus Publishing Company, New Delhi.

Chauhan, N.S. (1999): Medicinal and Aromatic Plants of Himachal Pradesh, Indus Publishing Company, New Delhi, PP 57-632.

Chauhan, N.S. (1987): Collection, Identification, Cultivation and Utilization of Medicinal and Aromatic Plants in H.P. Status Paper Presented to the Sub-Committee on Medicinal Plants Development in H.P. under the State Science, Technology and Environment Council, April 4, Shimla, PP 4-5.

Chauhan, N.S. (1999): Medicinal and Aromatic Plants of Himachal Pradesh, Indus Publishing Company, New Delhi, PP 14-44.

Chopra, I.C. and Handa, K.L. (1961): Review of Research on Indian Medicinal and Allied Plants, *Indian Council of Agricultural Research,* New Delhi, PP 27.

Chopra, R.N., Nayar, S.L. and Chopra, I.C. (1956): Glossary of Indian Medicinal Plants, *Council of Scientific and Industrial Research,* New Delhi.

Chowdhery, H.J. and Wadhwa, B.M. (1984): Flora of Himachal Pradesh, *Botanical Survey of India,* Howrah, I-III: PP 6-15.

Chuvieco, E. and Congalton, R. (1989): Application of Remote Sensing and Geographic Information System to Forest Fire Hazard Mapping, *Remote Sensing of Environment,* 29: PP 147-159.

Dalal, K.C., Joshi, P.P., Mandal, K. and Pandit, P.U. (1998): Strategy for Conservation and Availability of Medicinal Plants, Abstracts of National Symposium on Species, Medicinal and Aromatic Plants-Biodiversity, Conservation and Utilization (NS-MAP), Calicut, Kerala, PP 14.

Dasman, R.F. (1976): Environmental Conservation, Willy, New York, PP 220-240.

Dastur, J.F. (1962): Medicinal Plants of India and Pakistan, D.B Taraporevala Sons and Company, Pvt. Ltd., Bombay, PP 52-67.

Deeming, J.E., Burgan, R.L. and Cohen, J.D. (1978): The National Fire Danger Rating System, U.S. Department of Agriculture, Forest Service, Ogdan.

Dey, A.C. (1980): Indian Medicinal Plants Used in Ayurvedic Preparations, Bishen Singh, Mohendra Pal Singh, Dehradun, PP 139-150.

Dobriyal, R.M., Singh, G.S., Rao, K.S. and Saxena, K.G. (1997): Medicinal Plant Resources in Chhakinal Watershed in the North-Western Himalaya, *Journal of Herbs, Spices and Medicinal Plants,* 5(1): PP 15-27.

Dymock, W. et al. (1889-1891): Pharmacographica Indica, Thacker Sprink and Co., Calcutta, PP 54-66.

FAO (2001): Global Forest Resources Assessment 2000, Rome, Italy.

Gaur, R.D. and Singh, P.B. (1993): Ethonal-Medicinal Plants of Mandi District, Himachal Pradesh, *Bulletin of Ethnomedicobotany Research,* 14, PP 1-14.

Goldammer, J.G. and Seibert, B. (1990): The Impact of Droughts and Forest Fires on Tropical Lowland Rain Forest of East Kalimantan, Springer-Verlag, Berlin-Heidelberg, Germany, PP 11-31.

Goldammer, J.G. and Penafiel, S.R. (1990): Fire in the Pine-Grassland Biomes of Tropical and Subtropical Asia, Springer-Verlag, Berlin-Heidelberg, Germany, PP 44-62.

Gupta, R. (1964): Survey Record of Medicinal and Aromatic Plants of Chamba Forest Division of H.P., *Indian Forester,* 90: PP 454-468.

Gupta, R. (1971): Medicinal and Aromatic Plants of Bhadal Ranges, Chura Forest Division, Chamba District of H.P., *Journal of Bombay Natural History Society,* 68: PP 791-803.

Gupta, S.K (1986): Structure and Functioning of Natural and Modified Grassland Ecosystems in the Western Himalaya, Final Technical Report MBA/DOC, Govt. of India, PP 8-12.

Haines, H.H. (1921-1924): The Botany of Bihar and Orissa, Parts 6, London. Reprinted 1961, *Botonical Survey of India,* 1-3, Calcutta.

Hodges, A.W. (1995): Management Strategies for a Borehole Resin Production System in Slash Pine, Dissertation Submitted to the University of Florida, May, 1995, Gainesville, FI.

Hodges, A.W. and Willianms, G. (1993): Pine Gum in Bottle: A Sealed Collection Systm for Production of High Purity Oleoresin, Naval Stores Review 103: PP 2-8.

Holley, J. and Cherla, K. (1998): The Medicinal Plants Sector in India, Medicinal and Aromatic Plants Program in Asia (MAPPA), New Delhi, PP 91.

Hooker, J.D. (1872-1897): The Flora of British India, Reeve and Co. London.

Jadhav, R.N. (1988): An Evaluation of IRS-1A LISS Data for Forest Type Mapping and its Comparison with Landsat MSS Data, in Remote Sensing Application Using IRS-1A Data, Scientific Note, Space Application Centre, Ahmedabad, PP 53-62.

Jain, S.K. (1968): Medicinal Plants, National Book Trust, India, New Delhi, PP 1-216

Kaushik, P. and Dhiman, A.K. (2000): Medicinal Plants and Raw Drugs of India, Bishan Singh Mahendra Pal Singh, Dehradun.

Khory, R. et al. (1984): Meteria Medica of India and Their Therapeutics, Neeraj Publishing House, Delhi-52: PP 1-72.

Kirtikar, K.R. and Basu, B.D. (1975): (Reprint): Indian Medicinal Plants, Lalit Mohan Basu, 49, Leader Road, Allahabad, India, I-IV: PP 24-56.

Kumar, J. Singh, N.C. Shah and Rajan, V. (1997): Indian Medicinal and Aromatic Plants Facing Genetic Erosion, *Central Institute of Medicinal and Aromatic Plants,* Lucknow, PP 1-219.

Lal, P. (2000): National Forest Policy and Raw Material Supplies for Wood Industries in India, *Indian Forester*, 126 (4): PP 351-366.

Lange, D. (1997): Trade Figure for Botanical World-Wide, *Medicinal Plants Conservation,* 3: PP 16-17.

Laurie, A.P. (1910): Material and Painter's Craft Published G.T. Faulis and Co. Ltd., London, PP 21.

Latifovic, R., Fytas, K. and Paraszczak, J. (2005): Assessing Land Cover Change Resulting from Large Surface Mining Development, *International Journal of Applied Earth Observation and Geoinformation,* 7(1): PP 29-48.

Mahato, R.B. and Chaudhary, R.P. (2005): Ethnomedicinal Plants of Palpa District, Nepal, *Ethnobotany,* 17: PP 152-163.

Malhotra, K.C. et al. (1991): Role of Non Timber Forest Produce in Village Economy: A Household Survey in Jamboni Range, Midnapore District, West Bengal, PP 32-36.

MoEF (2002): National Forestry Action Programme, India, Government of India, New Delhi.

Negi, S.S. (1986): A Handbook of Forestry, International Book Distributors, Dehradun, India.

Nicholas, J.D. (1950): Adhesive for Metals Theory and Technology, Mcgraw Hill Book Company Inc, New Delhi, PP 52.

Odum, E.P. (1971): Fundamentals of Ecology, 3rded. W.B. Saunders Co. Philadelphia, PP 1-15.

Oza, G.M. (1972): On the Botanical Identify of Bhang, Charas, Judiciary and Society, *Indian Forester,* 98 (6): PP 349-355.

Paull, D. et al. (2006): Monitoring the Environmental Impact of Mining in Remote Locations through Remotely Sensed Data, Geocarto International, 21(1): PP 33-42.

Pirazizy, A.A. (1991): Environmental Sustainability, Dwindling Ecosystem Energetics and Human Sensibility: Assessment of Change in Temperate Forest, *The Fragile Environment, New Environment Series,* Ashish, New Delhi, PP 60-71.

Prakash, A. and Gupta, R. (1998): Land Use Mapping and Change Detection in a Coal Mining Area-A Case Study in the Jharia Coalfield, India, *International Journal of Remote Sensing,* 19(3): PP 391-410.

Pushpangadan, P. (1998): Bioresources and the Patent Regimes, Paper Presented at the International Seminar on Biotechnologies for Dry Land, Agriculture, July 16-18. Biotechnology Unit, IPE, Hyderabad.

Raghav, N. (1971): Pine Resin is Economic and Industrial Development of India, *Indian Forester,* 7, PP 11.

Rome, W.H. and Despain, D.G. (1989): The Yellow Stone Fires Scientific American, 261(5): PP 21-29.

Rigina, O. (1999): Monitoring of Forest Damage in the Kola Peninsula, Northern Russia Due to Smelting Industry, *The Science of the Total Environment,* 229: PP 147-163.

Rigina, O. (2002): Environmental Impact Assessment of the Mining and Concentration Activities in the Kola Peninsula, Russia, by Multidate Remote Sensing, *Environmental Monitoring and Assessment,* 75(1): PP 13.

Roy, M.M. (1999): Silvopastoral Systems, *Agroforestry Today,* 11(2): PP 24-26.

Roy, P.S., Das, K.K. and Naidu, K.S.M. (1991): Forest Cover and Landuse Mapping in Karbi Anglong and North Cachar Hills District of Assam Using Landsat MSS Data, *Journal of Indian Society of Remote Sensing,* 19(2): PP 112-123.

Samant, S.S., Dhar, U. and Palni, L.M.S. (1998): Medicinal Plamnts of Indian Himalaya: Diversity, Distribution Potential Values, Himvikas Publication No. 13 Gyanodaya Prakashan, Nainital, PP 163.

Sarin, M. (1995): Regenerating India's Forests: Reconciling Gender Equity with Joint Forest Management, in Susan Jockes Melissa Leach and Cathy Green (Eds,) Gender Relations and Environmental Control Changes, IDS Bulletic, 26(1): PP 83-91.

Sekhr, C., Rai, V. and Surendran, C. (1993): Price Regime Analysis, Marketing and Trade of Minor Forest Products-A Case Study, Center of Minor Forest Products: Dehradun.

Shank, M. (2009): Mapping Vegetation Change on a Reclaimed Surface Mine Using Quickbird, in: R. Barnhisel (Editor), National Meeting of the American Society of Mining and Reclamation, Revitalizing the Environment: Proven Solutions and Innovative Approaches, ASMR, Billings, MT, USA.

Shankar, D. and Majumdar, B. (1995): Non- Wood Forest Products Series No. 11. *Medicinal Plants for Conservation and Health Care,* FAO, Rome.

Sharma, D.D. (2005): Forests Economy and Environment, Kilaso Books, New Delhi.

Sharma, M. and Kaur, H. (1999): Additions to the Flora of Sirmour District, *Journal of Bombay Natural History Society,* 96: PP 93-97.

Simmons, J. et al. (2008): Forest to Reclaimed Mine Land Use Change Leads to Altered Ecosystem Structure and Function, *Ecological Applications,* 18(1): PP 104-118.

Singh, P.B. and Aswal, B.S. (1992): Medicinal Plants of Himachal Pradesh Used in Indian Pharmaceutical Industry, *Bulletin of Ethnomedicobotany Research,* 13: PP 172-208.

Sood, R.P., Kalia, N.K. and Sobti, S.N. (1982): Scope for Development of Phytochemical and Essential Oil Industry in Palampur Area (Kangra Valley), Nagarjun: PP 30-34.

State Forest Report (SFR) (2009): Forest Survey of India, Ministry of Forest and Environment, PP 199.

State Grazing Policy (1968): Report of the Grazing Advisory Committee on the Grazing Policy of Himachal Pradesh, PP15-16.

Tanaka, S., Kimura, H. and Suga, Y. (1983): Preparation of a 1:25,000 Landsat Map for Assessment of Burnt Area in Etayima Island, *International Journal of Remote Sensing,* 4(1): PP 17-31.

Tandon, V. (1997): The Status of Collection, Conservation, Trade and Potential for Growth in Sustainable Use of Major Medicinal Plant Species Found in the Great Himalayan National Park and its Environs in Kullu Distt. of Himachal Pradesh, Report Submitted to Wildlife Institute of India, Dehradun, PP 39.

Tansley, A.G. and Chipp, T.F. (1926): Aim and Methods in the Study of Vegetation, Published by British Empire Vegetation Committee and the Crown Agents for the Colonies, London, PP 140-151.

Townsend, P. et al. (2009): Changes in the Extent of Surface Mining and Reclamation In the Central Appalachians Detected Using A 1976-2006 Landsat Time Series, *Remote Sensing of Environment,* 113: PP 62-72.

Troups, R.S. (1921): The Silviculture of Indian Tree, VIII, Oxford University Press, Oxford, PP 1077.

Turner, B. et al. (2003): Illustrating the Coupled Human-Environment System for Vulnerability Analysis: Three Case Studies, *Proceedings of the National Academy of Sciences,* 100: PP 8080-8085.

Uniyal, M.R. and Chauhan, N.S. (1971): Medicinal Plants of Uhal Valley in Kangra Forest Division, Himachal Pradesh, *Indian Journal of Medical Research,* 6: PP 287-299.

Watt, G. (1972): A Dictionary of the Economic Products of India, *Periodical Experts,* Delhi, PP 486.

Working Plan of Forest (2014): Working Plan for the Forests of Renuka Forest Division, H.P. Govt., Forest Department.

Whyte, R.O. (1968): Land Livestock and Human Nutrition in India, *Praeger,* New York, PP 125-30.

CHAPTER - VI

LAND USE / LAND COVER PATTERN IN RENUKA FOREST DIVISION

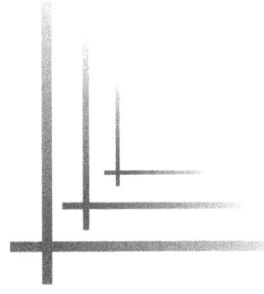

The fifth chapter was based on human impact on the physical environment that analyzed unscientific mining, exploitation of non-timber product, livestock grazing beside studying the people's perception about forest sector policies. The present chapter relay on the concept of land use and land cover pattern and classification, land use and land cover pattern and changes of the Renuka forest division and people's perception about land use and land cover pattern.

6.1 CONCEPT OF LAND USE / LAND COVER PATTERN

Human beings have been altering the face of the earth for the last few centuries but with the introduction of machines, the land cover of the earth has changed drastically in the last three centuries especially after the industrial revolution. The debate about the relationship between human population dynamics and the availability of natural resources dates back to more than 200 years when Malthus (1798) put forward his argument that population growth would eventually outstrip the production capacity of the land. It was only in the second half of the 20th century when the probability of the Malthusian projection seemed to be a reality, that sincere efforts to study the human population-environment relation were undertaken. The Scientific study and analysis of land use and land cover change involves a quantitative estimation of land use and land cover at a particular location and time. In this regard, remote sensing plays a major role in giving a synoptic view of the spatial extent of land use and land cover at a particular point of time.

The Human use of land resources gives rise to "land use" which varies with the purpose it serves, whether it be food production, provision of shelter, recreation, extraction and processing of materials, and the biophysical characteristics of the land itself. Tropical ecosystems are under continuous threat by organic and chemical pollution from agriculture and industries and the resultant degradation of the natural resources has taken on an alarming aspect (Benidick, 1999). In the developing countries, due to population pressure and in a bid to extract the maximum output from the available sources, the impact of degradation can be worse than in other countries and adversely affect the land cover of the region.

Land cover refers to the physical and biological cover over the surface of land, including water, vegetation, bare soil and/or artificial structures (Ellis, 2007). Land use, on the other hand, has a more complicated aspect as it involves social sciences and management principles and is defined as the social and economic purposes and contexts for and within which lands are managed. Although land use and land cover are frequently used as synonymous to each other yet, there is a very clear difference between the two (Gautam, 2004). While land cover signifies the spatial distribution of the different land cover classes on the earth's surface, and can be directly estimated qualitatively as well as quantitatively by remote sensing, land use and its changes require the integration of natural and social scientific methods to determine which human activities are occurring in different parts of the landscape, even when the land cover appears to be same (Lambin et al. 2001).

Land use and land cover change are perhaps the most prominent form of global environmental change as they occur at spatial and temporal scales and therefore the issue is relevant to our daily existence. Technically, land use and land cover change mean quantitative changes in areal extent (increase or decrease) of a given type of land use and land cover respectively. The land use and land cover change are a manifestation of forces both anthropogenic and environmental climate driven factors (Liu et al. 2009). The changes in land use in various spatial and temporal domains are the material expressions, and also indicate environmental and human dynamics and their interactions mediated by land availability (Lambin et al. 2001). Spatial data on land use and land cover in a region is a prerequisite to determine the qualitative and quantitative changes in land use and land cover. The technological advances in the field of remote sensing over the past few decades now enable repeated observations of the earth's surface (NAP, 2008). With the increase in sensor capability in terms of spatial resolution, spectral variability and temporal frequency, the minute changes on the earth's surface can be estimated more accurately.

The land use and land cover changes, apart from changing the physical dimension of the spatial extent of the land use and land cover classes, also influence many of the secondary processes which lead to the eventual degradation of the ecosystems of the earth (Dregne and Chow, 1992). The first and foremost, impact of land use and land cover changes is the reduction in vegetation cover. The loss of a vegetation cover, in turn, leads to many other deleterious effects on the environment, namely, loss of biodiversity, climate change, changes in radiative forcing, pollution of other natural ecosystems with a reduction in their quality, changes in hydrological regimes, and the list continues (Niyogi et al. 2009). The secondary impact of land use and land cover changes initiates a cascade of effects on the environment and this works in a loop to further influence land use and land cover changes.

6.1.1 Classification Criteria

A land use and land cover classification system, which can effectively employ orbital, and high-altitude remote sensor data should meet the following criteria (Anderson, 1971):

- The minimum level of interpretation accuracy in the identification of land use and land cover categories from remote sensor data should be at least 85 percent.
- The accuracy of interpretation for the several categories should be about equal.
- Repeatable or repetitive results should be obtainable from one interpreter to another and from one time of sensing to another.
- The classification system should be applicable over extensive areas.
- The categorization should permit vegetation and other types of land cover to be used as surrogates for activity.
- The classification system should be suitable for use with remote sensor data obtained at different times of the year.
- Effective use of subcategories that can be obtained from ground surveys or from the use of larger scale or enhanced remote sensor data should be possible.
- Aggregation of categories must be possible.
- Comparison with future land use data should be possible.
- Multiple uses of land should be recognized when possible.

Land use classification is the systematic arrangement of various types of available land on the basis of certain defined characteristics, mainly to identify and understand their fundamental utility. Thus land must be carefully presented and utilized so that it may fulfill our multifarious requirements. Conventionally, a nine-fold classification **(Fig. 6.1)** has been internationally adopted for easy comparability and scientific management of land. Land utilization units by directorate of economic and statistics New Delhi 1990 has also classified land into nine categories. This classification is also known as revenue classification. All the statistics related agriculture and other are collected at village level by conventional method and complied at district headquarters for publication. These statistics are used for national project planning and other development activities at local level (Gautam, 2002).

Nine Fold Land Use Classifications

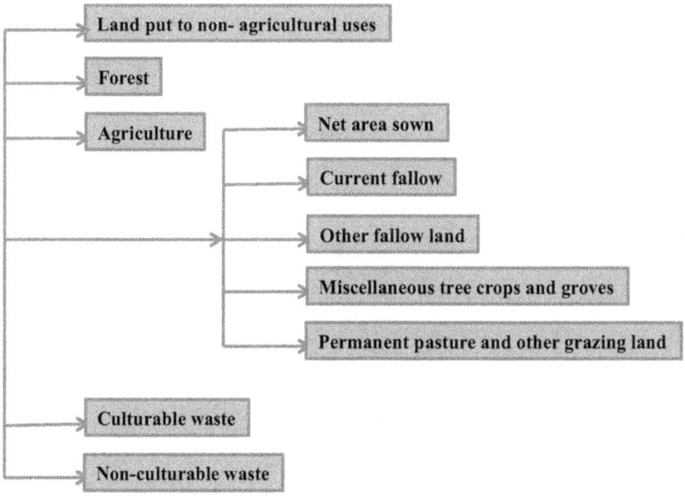

Fig. 6.1 Land use classification used by directorate of economics and statistics, Govt. of India

The directorate of land record Himachal Pradesh has also adopted this nine-fold land use classification. Similar to the national level, all the statistical related agriculture; land use classification and other in the state are collected at village level by conventional method and compiled at district headquarters for publication. In the present study four kinds land use classification has been used i.e. forest, agriculture, open land/ grass land/shrub land and water body.

6.2 LAND USE AND LAND COVER PATTERN IN THE RENUKA FOREST DIVISION

The principles of image classification are that a pixel is assigned to a class based on its feature vectors, by comparing it to predefined clusters in the feature space. Doing so far, all image filter results in a classified image. With the objective of converting the image data into thematic data with the most important characteristics of the vegetation in the area supervised classification was used. The spectral characteristics of the class identifying sample areas (training area) were defined. One requirement of supervised classification is to be familiarized with the area characteristics in case the field data has been used. For the land-use classification, Landsat TM, Landsat ETM+ and Resource Sat-II images were used. In order to have better accuracy supervised classification has used to prepare the land use map. The result shows that forest is major land use, which is followed by agricultural land, open land/grass land/shrub and other water. The land use status of different dated images is given below.

6.2.1 Land Use / Land Cover Classification of the Renuka Forest Division of 1972

For the classification of 1972 image, Landsat MSS satellite image was used. The land use land cover status is given in the **table 6.1** and also shown through LU/LC map **(Fig. 6.2)**. The analysis of Landsat image, data shows that forest occupies more than half of total land and it is the most important land use land cover of the study area. The agriculture occupies about 18 percent of the total geographical area while open land/grass land/shrub land consists of 19 percent, and occupies less than one percent is under the water bodies.

Table: 6.1 Land Use / Land Cover Pattern in Renuka Forest Division of 1972

Land Use Land Cover 1972		
	Status	
Land Category	Area in Sq.km.	% age
Forest	610	61.80
Agriculture	178	18.03
Open Land/Grass Land/Shrub Land	190	19.25
Water Body	9	0.91
Total Area in Sqkm.	987	100.00

Source: Data calculated by author from Landsat MSS Imagery

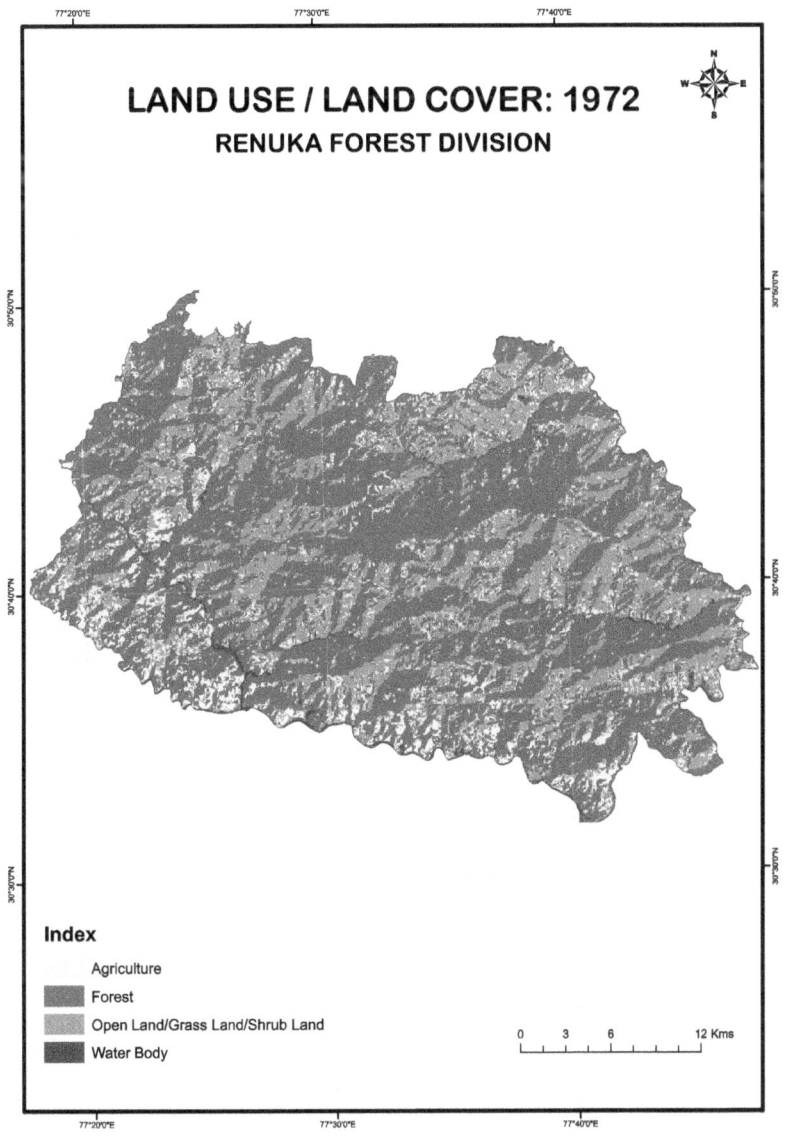

Fig. 6.2

6.2.2 Land Use / Land Cover Classification of the Renuka Forest Division of 1989

For the classification of 1989 image, Landsat TM satellite image was used. The land use land cover status is given in the **table 6.2**. The analysis of the table and map **(Fig. 6.3)** reveals that the forest still remains the major land use land cover consisting of more than half of total land. Agriculture area has declined to about 12 percent i.e. a sharp decline of about 6 percent. The open land/grass land/shrub land has increased to 23 percent. Similarly, area under water has also decreased to 0.8 percent.

Table: 6.2 Land Use / Land Cover Pattern in Renuka Forest Division of 1989

Land Use Land Cover 1989		
	Status	
Land Category	Area in Sq.km.	% age
Forest	634	64.24
Agriculture	115	11.65
Open Land/Grass Land/Shrub Land	230	23.30
Water Body	8	0.81
Total Area in Sqkm.	987	100.00

Source: Data calculated by author from Landsat TM Imagery

6.2.3 Land Use / Land Cover Classification of the Renuka Forest Division of 2001

For the classification of 2001 image, Landsat ETM+ satellite image was used. The land use land cover status is given in **table 6.3**. In order to appreciate the spatial changes, the land use land cover map has been prepared **(Fig. 6.4.)**. The thematic categories of map show that forest decreased upto 558 sq. km (56 percent) from the earlier 634 sq.km. in 1989. The area under agriculture decreased further to just 10 percent while; open land/grass land/shrub Land has increased to 32 percent of the total geographical area. There has been further decrease in the water bodies of this area as now they occupy just 0.7 percent of the study area.

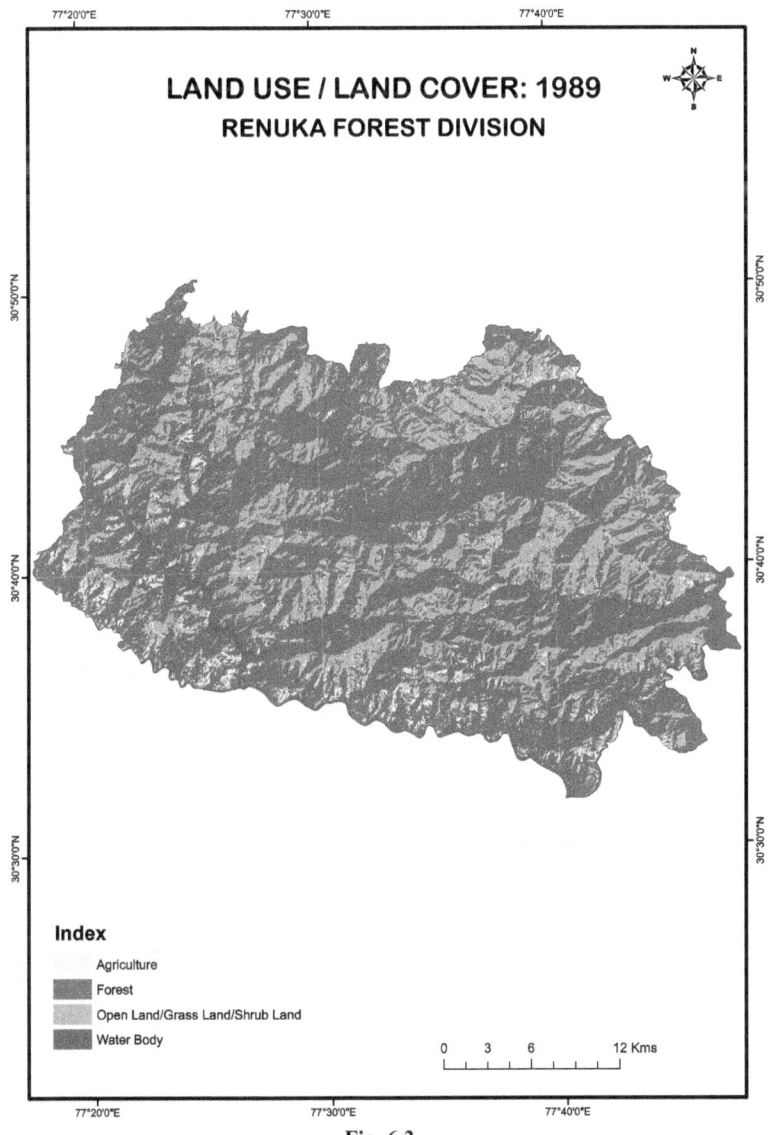

Fig. 6.3

Table: 6.3 Land Use / land cover pattern in Renuka forest division of 2001

Land Use Land Cover 2001		
	Status	
Land Category	Area in Sq.km.	% age
Forest	558	56.53
Agriculture	102	10.33
Open Land/Grass Land/Shrub Land	320	32.42
Water Body	7	0.71
Total Area in Sqkm.	987	100.00

Source: Data calculated by author from Landsat ETM+ Imagery

6.2.4 Land Use / Land Cover Classification of the Renuka Forest Division of 2011

For the classification of 2011 image, Resource Sat-II satellite image was used. The land use land cover statistics has been tabulated in **table 6.4** while the same has been shown through map **(Fig. 6.5)**. The analysis of this table and comparison with earlier LU/LC data shows that forest decreased from 56 percent to 55 percent of the study area. Agriculture increased of 19 percent, while open land/grass land/shrub land has declined to 19 percent. The water bodies in the study area have shrinked further and now occupy just 0.6 percent of the geographical area.

Table: 6.4 Land Use / Land Cover Pattern in Renuka Forest Division of 2011

Land Use Land Cover 2011		
	Status	
Land Category	Area in Sq.km.	% age
Forest	549	55.62
Agriculture	190	19.25
Open Land/Grass Land/Shrub Land	242	24.52
Water Body	6	0.61
Total Area in Sqkm.	987	100.00

Source: Data calculated by author from Resource Sat-II Imagery

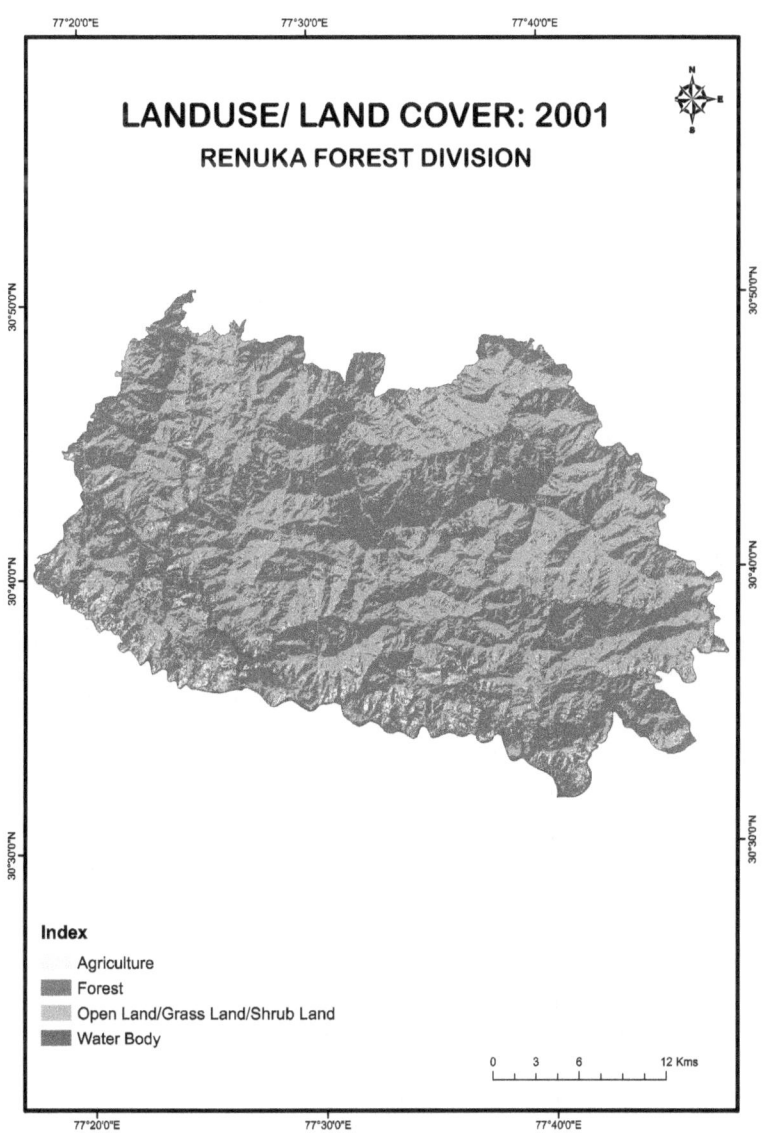

Fig 6.4

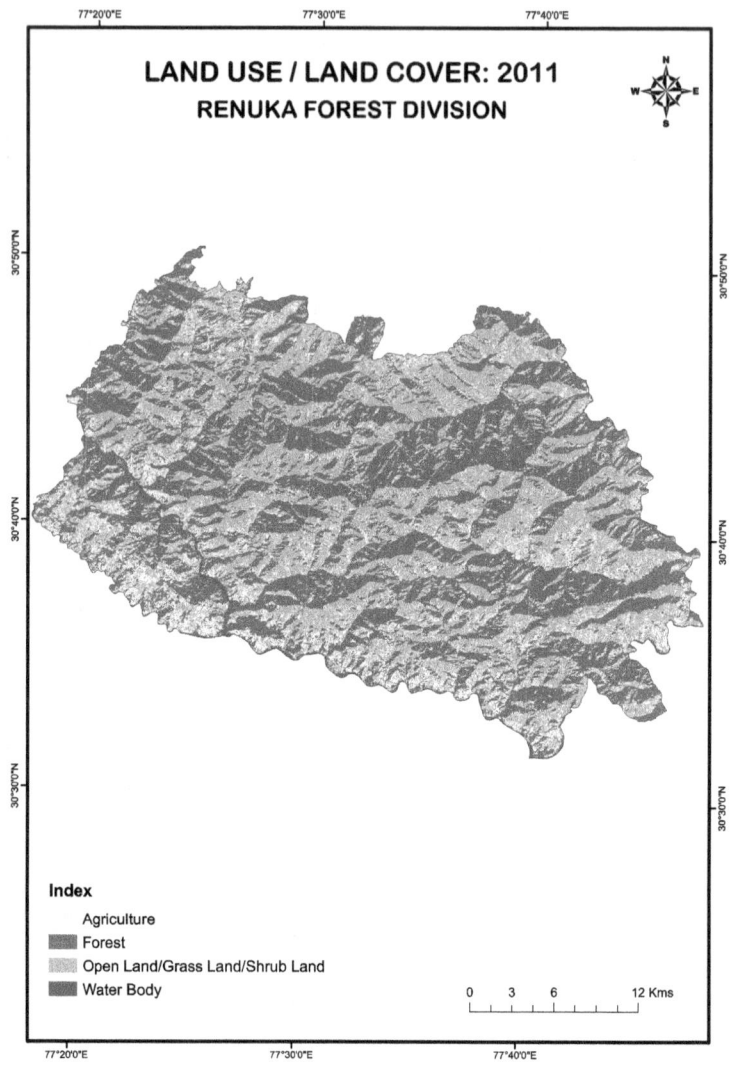

Fig 6.5

6.3 LAND USE AND LAND COVER CHANGES IN THE RENUKA FOREST DIVISION

Change detection refers to identifying differences in the state of an object or phenomenon by observing it at different times. Essentially, it includes the ability to quantify changes using multi temporal data sets one of the major applications of remotely sensed data obtained from earth-orbiting satellites is change detection because of repetitive coverage at short intervals and consistent image quality (Anderson, 1977). To study and identify the changes in land use and land cover pattern change detection is very important application. The change detection involves the use of multi-temporal data sets to discriminate areas of land use and land cover change between data of imaging. In the study area i.e. Renuka forest division change detection has been studied between 1972-1989, 1989-2001 and 2001-2011.

6.3.1 Land Use and Land Cover Changes in the Renuka Forest Division between 1972-1989

The comparison of land use/ land cover maps of 1972 and 1989 showed the considerable changes in forest, agriculture, open land/ grass land/ shrub land and water bodies **(Table 6.5)**. Between 1972 to 1989, forest has increased by 2 percent due to afforestation programmes carried out by forest department and also as a protection measures that facilitated natural growth. The afforestation programme carried out by forest department mainly in Renuka range (Ghataun), Kafota range (Tatiyana), Shillai Range (Shri Kyari, Chyali, Bhatnaul, Kota pab, Khatva, Milla, Jaswi, Lani, Baror, Dabar, Jarwa, Jakandon, Naipanjor, Tatwa Beyong), Sangrah Range (Daskana, Taikri, Panjah, Bhaltar, Lajwa, Arat, Ranphuwa, Uncha Tikkar) and Nohra Range (Manal, Chokar, Pipli, Bandal, Shilli, Bhangar, Bhangari, Nohra, Bhog, Charna, Ghandoori, Chunvi and Sail) have shown positive results. The study pertains to relatively small geographic units, hence, much variability is not expected within a short distance and some of changes might not be captured due to image resolution. There has been improvement in spatial resolution of Landsat satellites after the MSS sensor has been replaced with Thematic Mapper (TM). Therefore the artificial change in land use and land cover in the study area cannot be over ruled. The agriculture land use and water has decreased by 6 percent and 0.1 percent during this period. The farming practices were relatively more environmental friendly until 1990s. Whereas, the area under open land/grass land/shrub land registered an increase of 4 percent it may be due to agricultural fields have been left by the local people of the study area and they are converted to open land or in grass land etc. This has been shown through diagram **(Fig. 6.6)**.

Table: 6.5 Land Use /Land Cover Changes of the Renuka Forest Division between 1972-1989

Land Category	Land Use Land Cover 1972		Land Use Land Cover 1989		Land Use Land Cover Change	
	Area in Sq.km.	% age	Area in Sq.km.	% age	Area in Sq.km.	% age
Forest	610	61.80	634	64.24	24	2.43
Agriculture	178	18.03	115	11.65	-63	-6.38
Open Land/Grass Land/Shrub Land	190	19.25	230	23.30	40	4.05
Water Body	9	0.91	8	0.81	-1	-0.1
Total Area in Sqkm.	987	100	987	100	0	0

Source: Data calculated by author from satellite imageries, Landsat 1972 and 1989

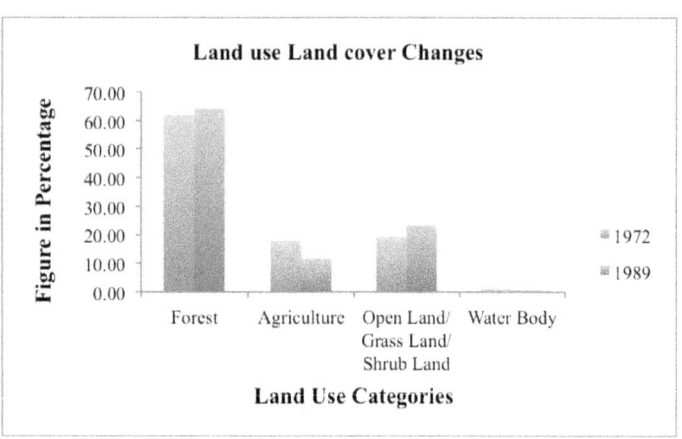

Fig. 6.6

6.3.2 Land Use and Land Cover Changes in the Renuka Forest Division between 1989-2001

The analysis of **table 6.6** shows that forest are of about 76 sq. km. has decreased which makes about 7 percent change in green cover. The principal cause of deforestation may be attributed to expansion of agricultural land vis-a-vis clearing of forest cover **(Photo plate 5.1)**. This happened more commonly in the low-lying areas having highest population density. Similarly agriculture area also decreased and brining a decrease of about 1 percent between 1989-2001. It has mainly happened due to the conversion of agriculture land into settlements and other rural development activities **(Photo plate 5.2)**. Open land/grass land/shrub land has increased substantially and registered a growth of 9 percent between 1989-2001. The water bodies have continued to decrease further. The shrinking water bodies are the results of transformation to various land uses as well as unabated construction activities. The broader view of changes in area between 1989 to 2001 has been also shown through diagram **(Fig. 6.7)**.

Photo plate 6.1 Deforestation for Expansion of Horticulture and Agriculture Activities at Banwani Area

Photo plate 6.2 Agriculture Land Converted into Rural Settlement at Trilor Dhar

Table: 6.6 Land Use / Land Cover Changes of the Renuka Forest Division between 1989-2001

	Land Use Land Cover 1989		Land Use Land Cover 2001		Land Use Land Cover Change	
Land Category	Area in Sq.km.	% age	Area in Sq.km.	% age	Area in Sq.km.	% age
Forest	634	64.24	558	56.53	-76	-7.70
Agriculture	115	11.65	102	10.33	-13	-1.32
Open Land/Grass Land/Shrub Land	230	23.30	320	32.42	90	9.12
Water Body	8	0.81	7	0.71	-1	-0.10
Total Area in Sqkm.	987	100.00	987	100.00	0	0.00

Source: Data calculated by author from satellite imageries, Landsat 1989 and 2001

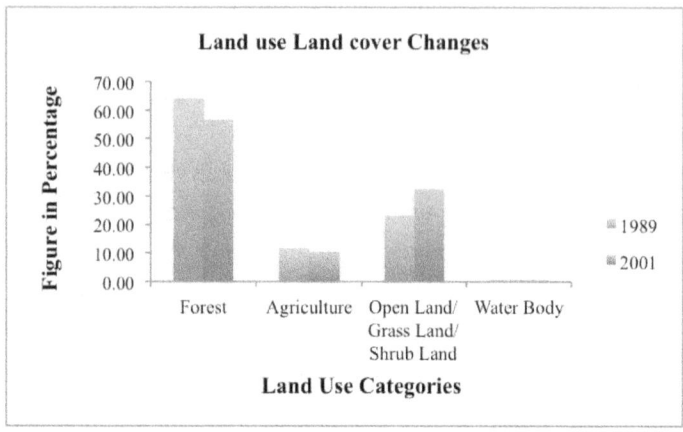

Fig. 6.7

6.3.3 Land Use and Land Cover Changes in the Renuka Forest Division between 2001-2011

The analysis of satellite data shows that area under forest has decreased by about 9 sq. km. this bringing a negative change in forest cover by about one percent. This decline in the forest cover is mainly due to the removal of a forests or stand of trees further paving the ways for non-forest uses of land e.g. grazing, settlement, road construction, installing hydropower's (Chandani, Manal and Timbi Hydro power projects **(Photo plate 6.3)** and mining activities **(Photo plate 6.5a,b)** in the study area. The encroachment **(Photo plate 6.4)** is a major cause of deforestation and forest degradation in Renuka forest division. The forests are depleting more because of illegal logging and conversion to non-forestry uses. The foremost reason of depletion of oak and pine forest was the clearing of land for agriculture and the over encroachment of forestland. The extensive field survey enabled the researcher to identify some of those areas where large-scale forest cover has been depleted and degraded. They are Charag, Ganu, Cho Boghar forests in Renuka Range, Khajuri, Jamna Pabar forests in Kafota Range, Balokothi, Koti Bonch, Kharkhan, Loja, Manal, Bandauli forests in Shillai Range and Jamal Nihog, Bhajond forests in Nohra Range. The area occupied by agriculture has increased by 8 percent in order to cater the growing demand as a result of rise in the human population. This has been very clearly reflected in the 7 percent decrease of open land/grass land/shrub land in the study area during the decade. **Table 6.7** and **figure 6.8** shows the change in area between 2001 to 2011.

Photo plate 6.3 Installed Hydropower at Manal   **Photo plate 6.4** Illegal Encroachment at Pabdhar

Photo plate 6.5a Mining Activities Leads to   **Photo plate 6.5b** Mining Activities at Barag
Deforestation at Pamta Area

Table: 6.7 Land Use /Land Cover Changes of the Renuka Forest Division between 2001-2011

	Land Use Land Cover 2001		Land Use Land Cover 2011		Land Use Land Cover Change	
Land Category	**Area in Sq.km.**	**% age**	**Area in Sq.km.**	**% age**	**Area in Sq.km.**	**% age**
Forest	558	56.53	549	55.62	-9	-0.91
Agriculture	102	10.33	190	19.25	88	8.92
Open Land/Grass Land/Shrub Land	320	32.42	242	24.52	-78	-7.90
Water Body	7	0.71	6	0.61	-1	-0.10
Total Area in Sqkm.	987	100.00	987	100.00	0	0.00

Source: Data calculated by author from satellite imageries, Landsat 2001 and Resource Sat 2011

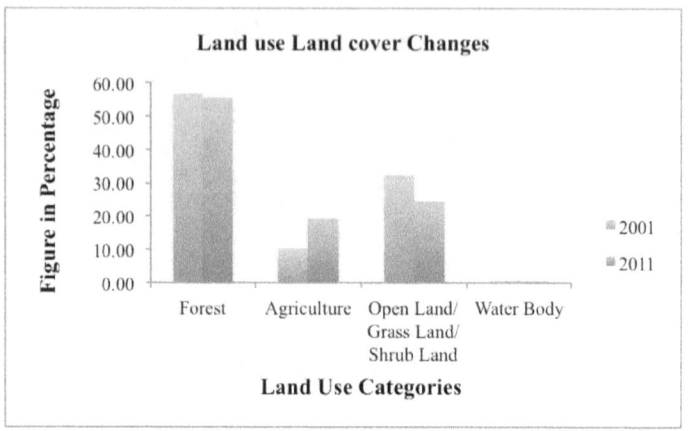

Fig. 6.8

6.4 PEOPLE'S PERCEPTION ABOUT LAND USE/ LAND COVER PATTERN

Land use involves the management and modification of natural environment or wideness into built environment such as fields, pastures, and settlements. It also has been defined as the arrangements, activities and inputs people undertake in a certain land cover type to produce, change or maintain it.

Table: 6.8 People's Opinion about Land Use of the Study Area

Types	No. of Respondents
Land under agricultural uses	60
Mining activities	10
Horticulture practices	20
Any other	10
Total	100

Source: Field Survey May, 2013

People's opinion about land use is an important tool to understand the matrix of reality and perception. **Table 6.8** shows that more than half of the respondents (60 percent) stated that land is under agricultural uses, 10 percent of people says that land use is under mining practices and one fifth of local believes that land use is under horticulture practices and another 10 percent finds other kind of land uses in the area under study.

Table: 6.9 People's Opinion about Land Cover of the Study Area

Opinion	No. of Respondents
Land under forest and vegetation	50
Barren and rocky land	30
Grass and grazing land	10
Miscellaneous	10
Total	100

Source: Field Survey May, 2013

Land cover is the physical material at the surface of the earth. Land covers include grass, trees, bare ground, water, etc. Land cover is distinct from land use despite the two terms often being used interchangeably. Land use is a description of how people utilize the land and socio-economic activity, urban and agricultural land uses are two of the most commonly known land use classes. Opinion about land cover brings out that around half of the respondents think that land cover is under forest and vegetation, whereas other 30 percent of people are of the view that land cover is barren and rocky. The **table 6.9** reveals that apart from the land under forest and vegetation and barren land cover categories, 10 percent says that grass and grazing land are the land cover. The remaining ten percent have said other miscellaneous things.

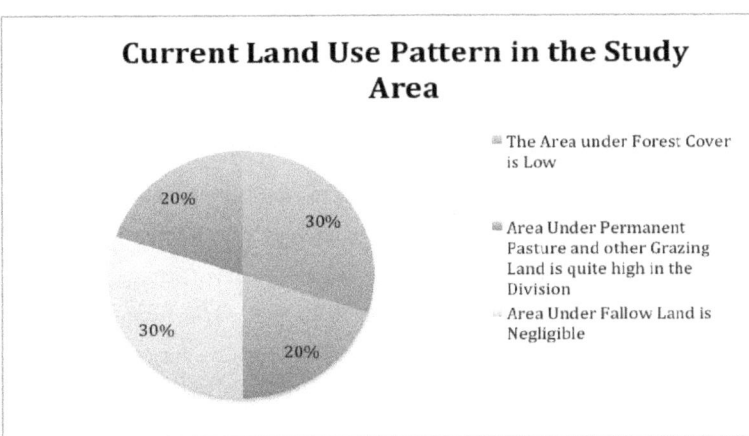

Source: Field Survey May, 2013

Fig. 6.9

According to the National Forest Policy, the minimum desired area, which is considered safe for a tropical country like India is about 33 per cent. As per broad policy recommendations, about 66 per cent of the area in the Himalayas and the Peninsular hills and 25 percent in the Great Plains should be under forests (Anon, 1988). The village grazing lands and even the permanent pastures require improvements in terms of soil and moisture conservation and increase in the nutritious and edible grass. **Fig. 6.9** reveals about current land use pattern in the Renuka forest division, 30 percent of people stated that the area under forest cover is low while 30 percent area under fallow land is negligible and one fifth of people stated that area under permanent pasture and other grazing land is quite high in the division and one fifth of land put to non-agricultural uses is quite low.

Table: 6.10 Forest Cover Changes in the Study Area

Type of Changes	No. of Respondents
Forest have depleted	20
Forest have degraded	25
Area under forest reduced	30
Forest area has gone up	25
Total	100

Source: Field Survey May, 2013

Forest degradation is the changes within the forest, which negatively affect the structure or function of the stand or site, and thereby lower the capacity to supply products or services. This takes different forms particularly in open forest formations deriving mainly from human activities such as overgrazing, overexploitation (for fuelwood or timber), repeated fires, or due to attacks by insects, diseases, and plant parasites. In most cases, degradation does not show as a decrease in the area of woody vegetation but rather as a gradual reduction of biomass, changes in species composition and soil degradation. Unsustainable logging practices can contribute to degradation if the extraction of mature trees is not accompanied with their regeneration or if the use of heavy machinery causes soil compaction or loss of productive forest area (Sharma, 2005). **Table 6.10** reveals about type of forest cover changes happened in the region, 30 percent of people stated that area under forest cover has reduced whereas one forth of the respondents stated that forests have degraded and one fifth of people also stated that forest area has gone up. Yet another 20 percent have opined that forests have depleted.

CONCLUSION

The geospatial techniques provide useful information for assessing the land use / land cover of any region. The change in landscape of the study area is mainly dominated by human in terms of substantial increase in area under agriculture, construction activities related to road, building and infrastructure development. These ventures have made a fast entry in the last two decades resulting in a large-scale land use/land cover changes in the study area. Summary, conclusions and policy imperatives shall be discussed in the last chapter.

REFERENCES

Aboyade, O. (2001): Geographic Information Systems: Application in Planning and Decision- Making Processes in Nigeria, Unpublished Paper Presented at the *Environmental and Technological Unit in the Development Policy Centre,* Ibadan.

Anderson, J.R. (1977): Land Use and Land Cover Changes: A Framework for Monitoring, *Journal of Research by the Geographical Survey,* 5: PP 143-153.

Anderson, J.R. (1971): Land Use Classification Schemes Used in Selected Recent Geographic Applications of Remote Sensing, *Photogrammetric Engineering and Remote Sensing,* PP 379-387.

Anderson, J.R., Hardy, E.E., and Roach, J.T. (1971): A Land Use Classification System for use with Remote-Sensor Data: U.S. Geological Survey, PP 16.

Anon, (1988): National Forest Policy 1988, Ministry of Environment and Forests, Government of India.

Asselman, N.E.M. and Middelkoop, H. (1995): Floodplain Sedimentation: Quantities, Patterns and Processes, *Earth Surface Processes and Landforms,* 20(6): PP 481-499.

Benedick, R.E. (1999): Tomorrow's Environment is Global, *Futures,* 31(10): PP 937-947.

Berlanga, C.A. and Ruiz, A. (2002): Land Use Mapping and Change Detection in the Coastal Zone of Northwest Mexico Using Remote Sensing Techniques, *Journal of Coastal Research,* 18(3): PP 514-522.

Coppin, P. et al. (2004): Digital Change Detection Methods in Ecosystem Monitoring: A Review, *International Journal of Remote Sensing,* (25)9: PP 1565-1596.

Dietz, T., Rosa E.A. and York, R. (2007): Driving the Human Ecological Footprint, *Frontiers in Ecology and Environment,* 5(1): PP 13-18.

Dregne, H.E. and Chou, N.T. (1992): Global Desertification Dimensions and Costs, In: Degradation and Restoration of Arid Lands, Lubbock: Texas Technical University.

El-Raey, M., Fouda, Y. and Gal, P. (2000): GIS for Environmental Assessment of the Impacts of Urban Encroachment on Rosetta Region, Egypt, *Environmental Monitoring and Assessment,* 60(2): PP 217-233.

Ellis E. (2007): Land Use and Land Cover Change, Encyclopedia of Earth.

Fung, T. and Ledrew, E. (1987): Application of Principal Components Analysis to Change Detection, *Photogrammetric Engineering and Remote Sensing,* 53(12): PP 1649-1658.

Gautam, N.C. (2002): Methodology for Land Use Planning: A Systematic Approach, Series-1 (First Edition), Center for Land Use Management, Hyderabad.

Gautam, N.C. (2004): National Land Use and Land Cover Classification, Series-3 (Second Edition), CLUMA Publication Hyderabad.

Hathout, S. (2002): The Use of GIS for Monitoring and Predicting Urban Growth in East and West St Paul, Winnipeg, Manitoba, Canada, *Journal of Environmental Management,* 66(3): PP 229-238.

Hudak, A.T. and Wessman, C.A. (1998): Textural Analysis of Historical Aerial Photography to Characterize Woody Plant Encroachment in South African Savanna, *Remote Sensing of Environment,* 66(3): PP 317-330.

Ingram, J.J. and Prochaska, D.D. (1972): Measuring Completeness of Coverage in the 1969 Census of Agriculture: American Statistical Association, *Business and Economy Sector Annual Management,* Montreal 1972, PP 199-215.

Jat, M.K., Garg, P.K. and Khare, D. (2008): Monitoring and Modelling of Urban Sprawl Using Remote Sensing and GIS Techniques, *International Journal of Applied Earth Observation and Geoinformation,* 10(1): PP 26-43.

Jensen, J.R. (1996): Introductory Digital Image Processing: A Remote Sensing Perspective, Prentice Hall, upper Saddle River, NJ, USA.

Lambin, E.F. et al. (2001): Our Emerging Understanding of the Causes of Land Use and Land Cover Change, *Global Environment Change,* 11: PP 261-269.

Lillesand, T.M. and Kiefer, R. (1993): Remote Sensing and Image Interpretation, Fifth Edition John Willey, New York.

Liu, S. et al. (2009): Quantifying the Spatial Details of Carbon Sequestration Potential and Performance, in *Science and Technology of Carbon Sequestration.*

Long, H., Wu, X., Wang, W. and Dong, G. (2008): Analysis of Urban-Rural Land Use Change During 1995-2006 and its Policy Dimensional Driving Forces in Chongqing, China, *Sensors,* 8(2): PP 681-699.

Maktav, D., Erbek, F.S. and Jurgens, C. (2005): Remote Sensing of Urban Areas, *International Journal of Remote Sensing,* 26(4): PP 655-659.

Malthus, T. (1798): An Essay on the Principle of Population, London.

Martinuzzi, S., Gould, W.A. and Gonzalez, O.M.R. (2007): Land Development, Land Use and Urban Sprawl in Puerto Rico Integrating Remote Sensing and Population Census Data, *Landscape and Urban Planning,* 79(4): PP 288-297.

Mas, J.F. (1999): Monitoring Land Cover Changes: A Comparison of Change Detection Techniques, *International Journal of Remote Sensing,* 20(1): PP 139-152.

NAP, (2008): Earth Observation from Space, www.nap.edu.

Navalgund, R.R. (2001): Remote Sensing, Resonance, 6(12): PP 51-60.

Niyogi D., Mahmood R. and Adegoke, J.O. (2009): Land Use and Land Cover Change and its Impacts on Weather and Climate, Boundary Layer Meteorology, 133(3): PP 297-298.

Ramakrishna, P.S. (1998): Sustainable Development, Climate Change and Tropical Rain Forest Landscape, Climatic Change, 39(2-3): PP 583-600.

Rao, S.S. (2008): Social Development In Indian Rural Communities: Adoption of Telecentres, *International Journal of Information Management,* 28(6): PP   474-482.

Roy, P.S. and Murthy, M.S.R. (2009): Efficient Land Use Planning and Policies Using Geospatial Inputs: an Indian Experience, in: Land Use Policy, Editors: A.C. Denman, O.M. Penrod, Nova Science Publishers, Inc.

Sharma, D.D. (2005): Forests Economy and Environment, Kilaso Books, New Delhi.

Sudhira, H.S., Ramachandra, T.V. and Jagdish, K.S. (2004): Urban Sprawl: Metrics, Dynamics and Modelling Using GIS, *International Journal of Applied Earth Observation and Geoinformation,* 5(1): PP 29-39.

TERI (1996): The Economic Impact of One Meter Sea Level Rise on Indian Coastline Methods and Case Studies, Report Submitted to the Ford Foundation.

Tziztiki, J.G.M., Jean, F.M. and Everett, A.H. (2012): Land Cover Mapping Applications with MODIS: A Literature Review, *International Journal of Digital Earth,* 5(1): PP 63-87.

Wackernagel, M. and Rees, W.E. (1996): Our Ecological Footprint: Reducing Human Impact on the Earth, New Society Publishers, PP 160.

Wackernagel, M. et al. (2002): Tracking the Ecological Overshoot of the Human Economy Proceedings of National Academy of Sciences USA, 99(14): PP 9266-9271.

Yeh, A.G.O. And Li, X. (1998): Principal Component Analysis of Stacked Multi-Temporal Images for fhe Monitoring of Rapid Urban Expansion in the Pearl River, *International Journal of Remote Sensing,* 19(8): PP 1501-1518.

CHAPTER - VII

SUMMARY, CONCLUSIONS AND POLICY IMPERATIVES

The foregoing chapter was about land use and land cover changes of the Renuka forest division, the present chapter proffer summary, conclusions and policy imperatives.

7.1 SUMMARY

The present study has been aimed to study the forest cover changes, forest sector policies and land use/ land cover pattern of the study area with a cohesive approach of remote sensing, geographic information system and analysis of socio-economic data. The underlying causes for the forest cover and land use/ land cover changes and modifications in the physical environment of the study area are multifarious. They include the encroachment by the villagers; open grazing inside the forest, and people's desire to expand forestland into cultivation land within the forest area. Renuka forest division of Sirmour district that forms the area of study lies between 30°31′11′′ and 30°52′16′′north latitudes and 77°17′34′′ and 77°47′38′′ east longitudes and extends over 987 sq. km. (Working Plan of Forest, 2014). The entire tract is mountainous and varies in elevation from 477 to 3647 metres above mean sea level. The slopes are generally steep to precipitous with deep stream and springs. The entire region of Renuka forest division falls within the catchments of Giri, Sainj and Tons rivers (District Gazetteer, 1996). Physiologically the tract forms the part of sub-Himalayan Zone and Lesser Himalayas. The sub-Himalayan zone extends to an altitude of 900 metre however the lesser Himalayan zone varies altitude from 900 to 2900 metres. The tract comprises of rocks ranging from Pre-cambrian to recent (Hayden, 1904). Soil is clayey loam, sandy clayey loam and clay in the study area. Temperature experiences both hot in summer as well as severe cold in winter and rainfall occurs maximum in summer and snow in winter in the higher altitude of the region.

Forests, which constitute an important resource and activities based on the utilization of this resource provide employment to a large number of people. But in the modern era this activity has substantially declined. The term forestry refers to obtaining of various types of products from forests. It includes not only the production of timber but also the activities of gathering of tree products. Production of timber is the most advanced among the forestry related activities, and gathering of various products is one of the oldest human occupations. While gathering is a form of primitive subsistence activity, lumbering or production of timber is a modern method of utilization of forest resources. Gathering of forest products is generally the way of life of the people with a low level of cultural and economic development (Singhal et al. 2003). Forests provide a multiplicity of environmental services. Foremost among these is the recharging of mountain aquifers, which sustain our rivers. They also conserve the soil, and prevent floods and drought. The forests provide habitat for wildlife and the ecological conditions for maintenance and natural evolution of genetic diversity of flora and fauna. In addition to it, they are the homes of traditional forest dependent communities and yield timber, fuel wood and other forest produce. The forests possesses immense potential for economic benefits, in particular for local communities, from sustainable eco-tourism. The principal direct cause of forest loss has been the conversion of forests to agriculture, settlements, infrastructure, and industry. In addition, commercial extraction of fuelwood, illegal felling, and grazing of cattle, has degraded forests. Forests are the most important resource of our globe. Naturally formed forests are found in such parts of globe where the factors of plant growth have been ideal for several centuries. For this reason, they are mostly found in regions of high rainfall and regions of high altitude.

The usefulness of forests is spread to commercial exploitation of forest for timber and other products, maintenance of birds and wildlife, conservation of ecological balance, prevention of soil erosion, etc. In different parts of world, there are evergreen forests, deciduous forests, coniferous forests scrub forest. Each one of them is different in respect of composition of species, atmospheric features of its location, density of plants and type of wildlife it has harboured. Forest resources are most important resources of our country useful in maintaining ecological balance, providing fire wood, providing raw materials to many industries, providing protection to wild animals and to conserve the soils. There are mainly four types of forests in India, evergreen forests are found in tropical tract of the country mainly concentrated in heavy rainfall hilly tracts of the country. Deciduous forests are found in plateau region of northern India and parts of central India. The coniferous forests are restricted to Himalayan and Sub-Himalayan regions. Scrub forests are distributed in all parts of the country. The per capita land availability getting reduced every year due to high population growth, forest area is lost at an alarming rate every year mainly to be converted into agricultural land (MoEF, 1999). Conservation of forest is certainly a necessity that requires to be addressed as a priority. For the survival of human beings, a holistic approach is required to be adopted with regard to protection of the plant kingdom as well as the wildlife. For the peaceful and mutually beneficial co-existence of all. Coming back to India, there are legislations abundant to deal with the situation by way of wildlife protection, forest conservation, etc. No doubt, the aims and object of such legislations are in tune with the call of the hour. Forests are being encroached by people who have been displaced from their original habitat for various reasons like construction of huge electricity generating dams, ethnic clashes, floods caused by breach of river embankments and dams etc. (Barbier, 1987).

The first scientific forest policy in India was formulated more than a century ago. The charter of Indian Forestry issued by the Government of India in 1855 heralded the beginning of forest conservancy in India. The first National Forest Policy of India published in 1894 (Kant and Cooke, 1999). The main objective of forest management was to promote the general well-being of the country. The influence of forest over the physical and climatic conditions of the country was quite well recognized. The second National Forest Policy was formulated after India's independence in 1952. The National Forest Policy of 1952, classified forests into protection forests, national forests, village forests and tree lands. The protection forests were meant for maintaining physical and climatic conditions, similar to preservation forests of the earlier policy. The National Forest Policy of India (1988) gave conservation orientation and a human face to forestry. The policy emphasized the protective role of forests in maintaining ecological balance and environmental stability (Anon, 1988). The 1988 national forest policy dictates that environmental conservation of forests and the meeting of subsistence needs of forest dependent communities should take precedence over commercial production. This is generally supported in Himachal Pradesh; there is a widespread feeling amongst senior government officials in the state that the primary purpose of forests should be for conservation and sustainable use. The state aims to bring 50 per cent of the area under tree cover, besides meeting all local requirements. At present there is no formal state forest policy document in Himachal Pradesh. However, the recent Forest Sector Review enabled substantial progress to be made towards the development of a new forest policy (Rabindranath and Joshi, 2006). The new Himachal Pradesh Forest Policy of 2005 states sustainable forest management as its chief goal and lists the following principles as its priorities: sustainable development, integration of natural resource management, decentralized governance, gender equity, and that forest policy should be more of a process- enabling it to be reviewed, adapted and revised as needed (GoHP, 2005).

The present research analysis also accentuates the role of remote sensing and geographic information system (GIS) in assessment of changes in forest cover, between 1972 to 2011, in the Renuka forest division. The trend of forest cover changes over the time span of 39 years was precisely analyzed. The study revealed that the forest cover was 610, 634, 558 and 549 sq.km in 1972, 1989, 2001 and 2011 respectively. It was noticed that forest cover has increased between 1972 and 1989, because of the implementation of various afforestation schemes by the forest department mainly in Renuka range (Ghataun), Kafota range (Tatiyana), Shillai Range (Shri Kyari, Chyali, Bhatnaul, Kota pab, Khatva, Milla, Jaswi, Lani, Baror, Dabar, Jarwa, Jakandon, Naipanjor, Tatwa Beyong), Sangrah Range (Daskana, Taikri, Panjah, Bhaltar, Lajwa, Arat, Ranphuwa, Uncha Tikkar) and Nohra Range (Manal, Chokar, Pipli, Bandal, Shilli, Bhangar, Bhangari, Nohra, Bhog, Charna, Ghandoori, Chunvi and Sail). The study also investigated that the forest cover has depleted and degraded between 1989 to 2011. The massive encroachment has reduced the forest to few relict pockets. Due to excessive biotic pressure, heavy exploitation for the purpose of timber, fuelwood extraction, grazing and other local uses, the forest cover has been reduced and many areas are degraded. The forest cover in the region incurred major losses during last decade due to increase in population. The field survey revealed that some of the areas have witnessed large-scale depletion and degradation of forest cover. The mention may be made of Cho Boghar forests in Renuka Range, Jamna Pabar forests in Kafota Range, Kharkhan, Loja, Manal, Bandauli forests in Shillai Range and Jamal Nihog, Bhajond forests in Nohra Range.

The study also tried to understand the human impact on the physical environment of the study area viz., impact of forest fire, unscientific mining or quarrying, exploitation of non-timber products and livestock grazing. In Renuka forest division, the fire is common in the Chil (Pinus roxburghii) areas and main cause of fire has been man and is due to negligence. The other reason of fire is the careless annual burning of the pastures by the locals wherein the fire escapes into the forests. The mining activities also causes the serious ecological problem of the study area such as degradation of vegetation cover, wild life, soil erosion, drying up of natural water springs etc. Similarly, the non-wood forest products occupy a significant position in the forest resources. From times immemorable the non-wood forest products like herbs and shrubs have been used for medicinal purposes. Further the study also includes the extraction of resin, one of the important non-wood forest products in the region. The present study tried to understand the different methods employed in the extraction of resin from Chir pine (Pinus roxburghii). An attempt has been made to analyze the impact of human on physical environment and they all resulted into deforestation and degradation of the forest lands of the study area. Forest degradation in the Renuka forest division is also largely due to ever increasing biotic pressure of man and their animals.

In the Renuka forest division land use and land cover change detection has been studied for the periods between 1972-1989, 1989-2001 and 2001-2011. The analysis of land use and land cover between 1972 and 1989 showed the considerable changes have occurred in forest, agriculture, open land/ grass land/ shrub land and water bodies. The change detection between 1972 to 1989 reveals that forest cover has increased by 2 percent due to afforestation programmes carried out by forest department and also as a protection measures that facilitated the natural growth. Another reason is that the study pertains to relatively small geographic units, hence, much variability is not expected within a short distance and some of changes might not be captured due to image resolution. There has been improvement in spatial resolution of Landsat satellites after the MSS sensor has been replaced with Thematic Mapper (TM). Therefore the artificial change in land use and land cover in the study area cannot be over ruled.

The analysis of satellite data for the year 1989 and 2001 shows that forest has decreased by 76 sq.km. which makes about 7 percent change in green cover. The principal cause of deforestation may be attributed to expansion of agricultural land. This happened more commonly in the low-lying areas having highest population density. Similarly agriculture area has also decreased due to the conversion of agriculture land into settlements and other rural development activities. The open land/grass land/shrub land has increased substantially and registered a growth of 9 percent between 1989-2001. Land use and land cover between 2001 to 2011 reveals that forest has decreased by about 9 sq.km. this bringing a negative change in forest cover by about one percent. This decline in the forest cover is mainly due to over grazing, expansion of rural settlement, road construction, installed hydropower's and mining activities in the study area. The extensive field survey enabled the researcher to identify some of those areas where large-scale forest cover has been depleted and degraded. They are Charag, Ganu, Cho Boghar forests in Renuka Range, Khajuri, Jamna Pabar forests in Kafota Range, Balokothi, Koti Bonch, Kharkhan, Loja, Manal, Bandauli forests in Shillai Range and Jamal Nihog, Bhajond forests in Nohra Range. The area occupied by agriculture has increased by 8 percent in order to cater the growing demand as a result of rise in the human population. This has been very clearly reflected in the 7 percent decrease of open land/grass land/shrub land in the study area during the decade.

The people's opinion about the forest cover and land use/ land cover changes and forest sector polices has been also taken into consideration. They also supported the above reasons for depletion and degradation of the forest cover of the study area.

7.2 CONCLUSIONS

Forest, the renewable natural resource is the backbone of the human civilization. It is essential in maintaining environment stability and plays a vital role in the ecological balances by providing habitat for flora and fauna, anchors soil and tames climate. Forests, which serve man multi-dimensionally to meet his basic needs and wants in the present day world; are the victims of man's greed. The unscientific use and exploitation of this natural resource has given birth to many environmental and ecological problems. The man who is not only destroying this natural resource by felling and removing trees and extracting other forest products unscientifically, but also constantly encroaching upon the forest land to satisfy the hunger for land. After extensive studies on human interaction with forest following conclusions have been drawn:

- The forest cover has increased between 1972 to 1989, due to efforts were underway to restore and rehabilitate degraded areas by bringing them under massive afforestation, social forestry and fuelwood/fodder development programs. The forest cover has changed drastically in the study area between 1989 to 2011 due to illegal encroachment, excessive biotic pressure, resulting heavy exploitation for the purpose of timber, fuelwood extraction and other local uses. This conclusion proved, first hypothesis of the research work.

• The impact of human on physical environment i.e. forest fire, unscientific mining, exploitation of non-timber products and livestock grazing leads to the depletion and degradation of forests cover which has left visible imprints on the physical environment of the study area and support the second hypothesis of the study.

• Between 1972 to 1989, forest cover has increased due to the area under open land/grass land/shrub land registered an increase, this may be due to the agricultural fields of the study area have been uninhabited by the locals and they are converted into forest land or grass land. Between 1989, 2001 and 2011 forest cover has decreased, the principal cause of deforestation was expansion in agricultural activities, over grazing, expansion in human settlement, road construction, installed hydropower's and mining practices in the study area. In earlier agriculture has also decreased because of some amount of agriculture land converted into settlements and other rural development activities. Between 1989, 2001 and 2011 agriculture has increased in order to cater the growing demand as a result of rise in the human population. This growing population pressure has been degraded the forests of the study area and verified the third hypothesis of the research.

• The water bodies have continued to decrease from 1972 to 2011. The shrinking water bodies are the results of transformation to various land uses as well as constant construction activities. Water resources specially in mining areas have been dried up and over utilization of forest also results reduction in the surface water supply in the springs, streams, wells and decrease of underground water reservoirs associated with attenuation of water table.

7.3 POLICY IMPERATIVES

After the detailed study, analysis of satellite data, field observations and on the basis of people's opinion in the study area, following are some of the policy imperatives or suggestions that can be made in order to have sustainable development of forests of the Renuka forest division.

➢ The forests in the study region are highly vulnerable to forest fires and this cause high loss of the forest wealth of the study area in every year. There is a need to apply both structural and non-structural measures to control the forest fire and a special task force should be established to mitigate and manage the menace.

➢ The open grazing system is causing a serious damage to the soil and vegetation of the study area. There is a need to reduce uncontrolled grazing and encourage the stall feeding so that the damage done to the environment by these animals is stopped.

➢ The area under forest cover has decreased over the period of time. There is a need to increase the forest cover through effective implementation of afforestation programmes.

➢ There is an urgent need to conserve the water resources specially in mining areas, since there were many incidents of drying up of natural water springs in the study area. This requires an active participation of local people.

➢ Efforts should be made to rehabilitate the mined areas in the study region. And proper legislation may be made to collaborate and restore the vegetation in the mined areas.

➢ The nomadic tribes namely Gujjars and Kinnauras and Chambiyal communities migrate along with the heard of their cattle to the high reaches of the study area during summer and return to the lower hills during winter. The seasonal movement leads to degradation and destruction of pasturelands and forests, which have a great impact on soil erosion. Efforts should be made by the government to rehabilitate them in permanent settlements, by providing them with free houses, opening schools and dispensaries for them and stall-feeding arrangement for their cattle.

➢ The *Panchayats* need to be strengthened so that they can monitor forest development, check illegal encroachment and destruction, and remove bottlenecks and defects of a centralized administrative system.

> A holistic approach is required to reduce the dependence of communities on forests, by providing them with direct opportunities to encouraging the use of locally available and renewable energy sources e.g. solar, wind and hydel energy sources.

> There is a need for continuous monitoring of changes in the forest cover, besides examining the impact of these changes on environment. The vulnerable forested areas require a special attention in this context.

The human interaction with forest are complex both in space and time. The all possible efforts have been made to study. The human and forest interaction in the Renuka forest division using geospatial techniques and socio economic investigation. The conclusions have been carefully drawn yet the evolving developmental activities and advent of new occupational structure affect the study area at large. The environment is at peril and the age old human forest interaction is also changing which needs to be continuously observed and analyzed. Therefore this is not the end but the beginning.

REFERENCES

Anon, (1988): National Forest Policy 1988, Ministry of Environment and Forests, Government of India.

Barbier, E. (1987): The Concept of Sustainable Economic Development, *Environmental Conservation,* 14(2): PP 101-110.

Chun, J. (2014): A Legal Approach to Induce the Traditional Knowledge of Forest Resources, *Forest Policy and Economics,* 38: PP 40-45.

Clough, J.D. (1968): Concept of Management Science, Prentice-Hall of India, New Delhi, PP 2.

District Gazetteer (1996): Gazetteer of the Sirmour District, Indus Publishing Company.

Gohp (2005): Forest Sector Policy and Strategy, Himachal Pradesh Forest Department, Government of Himachal Pradesh, Shimla.

Hayden, H.H. (1904): The Geology of Spiti with Parts of Bashar and Rupshu. Mem. Geological Survey India: 36(1).

Kant, S. and Cooke, R. (1999): Cultivating Peace: Conflict and Collaboration in Natural Resource Management in D. Buckles (Ed.), Cultivating Peace: Conflict and Collaboration in Natural Resource Management, Washington, IDRC- World Bank Institute, PP 81-97.

MoEF (1999): National Forestry Action Programme, India, Government of India, New Delhi.

Rabindranath, N.H. and Joshi, N.V. (2006): Impact of Climate Change on Forests in India, *Current Science,* 90(3): PP 354-361.

Singhal, R.M. et al. (2003): Forests and Forestry Research in India, *Tropical Ecology,* 44(1): PP 55-61.

Taylor, F.W. (1948): The Principle of Scientific Management, Harper and Row, New York, PP 36-40.

Working Plan of Forest (2014): Working Plan for the Forests of Renuka Forest Division, H.P. Govt., Forest Department.

BIBLIOGRAPHY

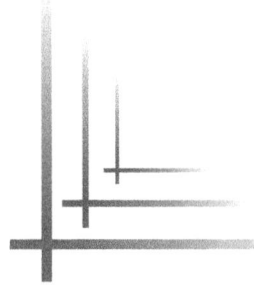

Abbas, I. et al. (2010): Mapping Land Use Land Cover and Change Detection in Kafur Local Government, Katsina, Nigeria (1995-2008), Using Remote Sensing and GIS, *Research Journal of Environmental and Earth Sciences,* 2(1): PP 6-12.

Abd El-Kawy, O.R. et al. (2011): Land Use and Land Cover Change Detection in the Western Nile Delta of Egypt Using Remote Sensing Data, *Applied Geography,* 31(2): PP 483-494.

Achard, F. et al. (2002): Determination of Deforestation Rates of the World's Humid Tropical Forests, *Science Magazine,* 297(55): PP 999-1002.

Adeniyi, P.O. and Omojola, A. (1999): Land Use Land Cover Change Evaluation in Sokoto-Rima Basin of North Western Nigeria Based on Archival of the Environment (AARSE) on Geoinformation Technology Applications for Resource and Environmental Management in Africa, PP 143-172.

Afify, H.A. (2011): Evaluation of Change Detection Techniques for Monitoring Land Cover Changes: A Case Study in New Burg El-Arab Area, *Alexandria Engineering Journal,* 50(2): PP 187-195.

Agnihotri, Y., Dubey, L.N. and Dyal, S.K.N.(1985): Effect of Vegetation Cover on Runoff from a Watershed in Shiwalik Foot Hills, *Indian Journal of Soil Conservation,* 13(1).

Aguirre, J. et al. (2012): Optimizing Land Cover Classification Accuracy for Change Detection, a Combined Pixel-Based and Object-Based Approach in a Mountainous Area in Mexico, *Applied Geography,*34: PP 29-37.

Allen, T.R. and Kupfer, J.A. (2000): Application of Spherical Statistics to Change Vector Analysis of Landsat Data: Southern Appalachian Spruce-Fir Forests, *Remote Sensing of Environment,* 74(3): PP 482-493.

Alphan, H. (2011): Comparing the Utility of Image Algebra Operations for Characterizing Landscape Changes: The Case of the Mediterranean Coast, *Journal of Environmental Management,* 92(11): PP 2961-2971.

Anand, V.K. (1982): A Note on the Structural Features of the Jutogh Formation in the Kullu Himalaya, Mandi and Kangra Distt. Himachal Pradesh, Geological Survey India, PP 41.

Anderson, et al. (1976): A Land Use and Land Cover Classification System for Use with Remote Sensor Data, *Geological Survey Professional,* U.S. Government Printing Office, Washington, D.C. PP 28.

Anderson, J.R. Hardy, E.E., Roach, J.I. and Winer, R.E. (1976): A Land Use and Land Cover classification System for Use with Remote Sense Data, *Geological Survey Professional Paper,* U.K.

Annual Report (2012-13): Department of Animal Husbandry, *Dairying and Fisheries, Ministry of Agriculture,* Govt. of India.

Anon, (1988): National Forest Policy 1988, Ministry of Environment and Forests, Government of India.

Aswal, B.S. and Mehrotra, B.N. (1994): Flora of Lahaul-Spiti: A Cold Desert in North-West Himalayas. Bishen Singh, Mahender Pal Singh, Dehra Dun and Periodical Express, Delhi, PP 761.

Awasthi, K.D. (2004): Land Use Change Effects on Soil Degradation, Carbon and Nutrient Stocks and Greenhouse Gas Emission in Mountain Watershed, Ph.d Thesis, Agriculture University, Norway.

Awasthi, K.D. et al. (2005): Analysis of Land Use Structure in Two Mountain Watersheds of Nepal Using FRAGSTATS, *Forestry: A Journal of Forestry Nepal,* 13(6): PP 495-513.

Awasthi, K.D., Sitaula, B.K., Singh, B.R. and Bajracharya, R.M. (2002): Land Use Change in Two Nepal's Watershed: GIS and Geomorphometric Analysis, Land Degradation and Development, 13: PP 495-513.

Bahuguna, V. K. (2002): Forest Policy Initiatives in India over the Last Few Years, in the Proceedings of The Forest Policy Workshop, 22-24 January, 2002, Kuala Lumpur.

Banko, G. (1998): A Review of Assessing the Accuracy of Classifications of Remotely Sensed Data and of Methods Including Remote Sensing Data in Forest Inventory, *International Institute for Applied Systems Analysis,* Luxemburg.

Barbier, E. (1987): The Concept of Sustainable Economic Development, *Environmental Conservation,* 14(2): PP 101-110.

Baronti, S. et al. (1994): Principal Component Analysis for Change Detection on Polarimetric Multitemporal SAR Data, *Geoscience and Remote Sensing Symposium, Surface and Atmospheric Remote Sensing: Technologies, Data Analysis and Interpretation.*

Bazzaz, F.A. (1975): Plant Species Diversity in Old Field Successional Ecosystems in Southern Illinois, *Ecology,* 56: PP 485-488.

Benediktsson, J. and Sveinsson, J. (1997): Feature Extraction for Multisource Data Classification with Artificial Neural Networks, *International Journal of Remote Sensing,* 18(4): PP 727-740.

Bennet, H.H. (1955): Elements of Soil Conservation, McGraw-Hill, New York.

Berberoglu, S. and Akin, A. (2009): Assessing Different Remote Sensing Techniques to Detect Land Use/Cover Changes in the Eastern Mediterranean, *International Journal of Applied Earth Observation and Geoinformation*, 11(1): PP 46-53.

Bharti, R.R., Ishwar, D.R., Adhikar, B.S. and Rawat, G.S. (2011): Timberline Change Detection Using Topographic Map and Satellite Imagery: A Critique, *Tropical Ecology*, 52(1): PP 133-137.

Bhati, J.P. (1979): Development Strategies in Himachal Pradesh, Mountain Farming

Bhattacharya, P.K. (1990): Social Forestry a Step Towards Environmental Change, Khana Publishers, New Delhi.

Bhattrai, K. and Dennis, C. (2007): Evaluating Land Use Dynamics and Forest Cover Change in Nepal's Bara District (1973-2003), *Journal Human Ecology*, PP 81-95.

Birendra, K.C. and Shin, N. (2006): Refugee Impact on Collective Management of Forest Resources: A Case Study of Bhutanese Refugees in Nepal's Eastern Terai Region, *Journal of Forest Research*, 11(5): PP 305-311.

Bliss, L.C. (1966): Plant Productivity in Alpine Micro-Environment on Mountain Washington, New Hampshire, *Ecology Monograph*, 36: PP 125-135.

Borgoyary, M., Saigal, S. and Peters, N. (2005): Participatory Forest Management in India: A Review of Policies and Implementation, Working Paper No.1 Overseas Development Group, University of East Anglia, Norwich, UK.

Borrough, P.A. (1986): Principles of Geographic Information System for Land Resources Assessment, Oxford, Clarendon Press, PP 193

Brent, C.J. and Peter, R.B. (2014): Drivers of Change in Landholder Capacity to Manage Natural Resources, *Journal of Natural Resources Policy Research*, 6(1): PP 1-26.

Brink, A.B. and Eva, H.D. (2009): Monitoring 25 Years of Land Cover Change Dynamics in Africa: A Sample Based Remote Sensing Approach, *Applied Geography*, 29(4): PP 501-512.

Brondizio, E. et al. (1994): Land Use Change in the Amazon Estuary: Patterns of Caboclo Settlement and Landscape Management, *Human Ecology*, 22(3): PP 249-278.

Buchy, M. (1995): The British Colonial Forest Policies in South India, A Maladapted Policy, in Yvon Chatelin and Christophe Bonneuil (Ed) Nature and Environment, Orstom Editions, Paris.

Byron, N. and Arnold, J.E.M. (1999): What Futures for the People of the Tropical Forests, *World Development*, 27(5): PP 789-805.

Castellana, L. et al. (2007): A Composed Supervised/Unsupervised Approach to Improve Change Detection from Remote Sensing, *Pattern Recognition Letters*, 28(4): PP 405-413.

Census of India (2011): Census of India, *Ministry of Home Affairs*, Govt. of India.

Chadha, S.K. (1989): Environmental Holocaust in Himalayan, Ashish Publishing House, New Delhi.

Champion, H.G and Seth, S.K. (1968): A revised Survey of forest Types in India, The Manager of Publications, Government of India, New Delhi.

Champion, H.G. and Seth, S.K. (1968): A Revised Survey of the Forest Types of India, *The Manager of Publication*, New Delhi.

Chaturvedi, A.N. (1981): Big Forest Loss Every Minute, *The Daily National Herald*, Lucknow. Oct 19, 1981.

Chaturvedi, O.P. and Singh, J.S. (1983): Estimation of Biomass and Biomass Production of Pinus Roxburghii Tree Growing in All Aged Natural Forests, *Canadian Journal of Forest Research*, 12(3): PP 632-640.

Chavez, P.S. and MacKinnon, D.J. (1994): Automatic Detection of Vegetation Changes in the Southwestern United States Using Remotely Sensed Images, *Photogrammetric Engineering and Remote Sensing*, 60(5): PP 571-583.

Cheng, S. et al. (1998): Deforestation and Degradation of Natural Resources in Ethiopia: Forest Management Implications From A Case Study in the Belete-Gera Forest, *Journal of Forest Research*, 3(4): PP 199-204.

Chowdhary, S. (1985): Environment and Resources of Tropical and Temperate Forests of India, Dehradun International Book Distributors.

Christensen, E.J. et al. (1998): Aircraft MSS Data Registration and Vegetation Classification for Wetland Change Detection, *International Journal of Remote Sensing*, 9(1): PP 23-38.

Chun, J. (2014): A Legal Approach to Induce the Traditional Knowledge of Forest Resources, *Forest Policy and Economics*, 38: PP 40-45.

Chytry, M. (2001): Phytosociological Data Give Biased Estimates of Species Richness, *Journal of Vegetation Science*, 12(3): PP 439-444.

Civco, D.L. et al. (2002): A Comparison of Land Use and Land Cover Change Detection Methods, *Proceedings of the 2002 ASPRS Annual Convention*, Washington DC.

Clark, A.N. (2003): Dictionay of Geography, Penguin Books.

Colwell, R.N. (1983): Manual of Remote Sensing (2nd ed), 1, *American Society for Photogrammetry and Remote Sensing,* Falls Church, Virginia.

Comber, A. et al. (2012): Spatial Analysis of Remote Sensing Image Classification Accuracy, *Remote Sensing of Environment,* 127: PP 237-246.

Conese, C. and Maselli, F. (1992): Use of Error Matrices to Improve Area Estimates with Maximum Likelihood Classification Procedures, *Remote Sensing of Environment,* 40(2): PP 113-124.

Congalton, R.G. (1991): A Review of Assessing the Accuracy of Classifications of Remotely Sensed Data, *Remote Sensing of Environment,* 37(1): PP 35-46.

Congalton, R.G. and Green, K. (2008): Assessing the Accuracy of Remotely Sensed Data: Principles and Practices, 2th Edition, CRC Press, Boca Raton.

Congalton, R.G. et al. (2001): Quality Assurance and Accuracy Assessment of Information Derived from Remotely Sensed Data, In: J. D. Bossler, Ed., *Manual of Geospatial Science and Technology,* CRC Press, London, PP 349-361.

Coppin, P. and Bauer, M. (1996): Digital Change Detection in Forest Ecosystems with Remote Sensing Imagery, *Remote Sensing Reviews,* 13: PP 207-234.

Coppin, P., Jonckheere, I., Nackaerts, K. and Muys, B. (2001): Digital Change Detection Methods in Ecosystem Monitoring, *International Journal of Remote Sensing,* 25(9): PP 1565-1575.

Coppin, P.P. and Bauer, M.E. (1994): Processing of Multitemporal Landsat TM Imagery to Optimize Extraction of Forest Cover Change Features. *IEEE Transactions of Geoscience and Remote Sensing,* 32(4): PP 918-927.

Damayanti, M.O.E. et al. (2004): Institutional Aspect of Protected Areas and Local People's Involvement into the Management: A Case Study in Periyal Tiger Reserve, Kerala, India, Presented at the Conference of Japan Forestry Economic Society, Tsukuba, 21, November, 2004.

Daniel, et al. (2002): A Comparison of Land Use and Land Cover Change Detection Methods, ASPRS-ACSM Annual Conference and FIG XXII Congress, PP 2.

Dansereau, P. (1960): Origin and Growth of Plant Communities, In: Growth in Living Systems Proceedings of International Symposium Held at Purdue University, Basic Book Inc., New York, PP 567-603.

Dasmann, R.F. (1968): Environment Conservation, John Willey and Sons, Inc. New York.

Desai, V. (1991): Forest Management in India- Issues and Problem, Himalaya Publishing House, Bombay.

Desclee, B. (2007): Automated Object-Based Change Detection for Forest Monitoring by Satellite Remote Sensing: Applications in Temperate and Tropical Regions, Ph.D Thesis.

Desclee, B., Bogaert, P. and Defourny, P. (2006): Forest Change Detection by Statistical Object-Based Method, *Remote Sensing of Environment,* 102: PP 1-11.

Dikshit, K.R. (1991): Environment, Forest Ecology and Man-In Western Ghats, Rawat Publications, Jaipur.

Dimyati (1995): An Analysis of Land Use and Land Cover Change Using the Combination of MSS Landsat and Land Use Map-A Case Study of Yogyakarta, Indonesia, *International Journal of Remote Sensing,* 17(5): PP 931-944.

Dimyati, M.U.H. et al. (1996): An Analysis of Land Use/Cover Change in Indonesia, *International Journal of Remote Sensing,* 17(5): PP 931-944.

District Gazetteer (1996): Gazetteer of the Sirmour District, Indus Publishing Company.

Dongsheng, L. et al. (2004): Forest Resources and Environment in China, *Journal of Forest Research,* 9(4): PP 307-312.

Dwivedi, A.P. (1980): Forestry in India, Dehradun, India, Jugal Kishore and Company, New Delhi.

Dwivedi, A.P. (1993): A Text Book of Silviculture, International Book Distributors, Dehradun, PP 975.

Dwivedi, A.P. (1993): Forests, the Ecological Ramifications, Natraj Publishers, Dehradun, PP 14-15.

Efe, R. et al. (2012): Land Use and Land Cover Change Detection in Karinca River Catchment (NW Turkey) Using GIS and RS Techniques, *Journal of Environmental Biology,* 33(2): PP 439-447.

Ellum, D.S. (2009): Floristic Diversity in Managed Forests: Demography and Physiology of Understory Plants Following Disturbance in Southern New England Forests, *Journal of Sustainable Forestry,* 28: PP 132-151.

EOSAT (1992): Landsat TM Classification International Georgia Wetlands in EOSAT Data User Notes, EOSAT Company, Lanham, MD, (7)1.

EOSAT (1994): EOSAT's Statewide Purchase Plan Keeps South Carolina Residents in the Know, in EOSAT Notes, EOSAT Company Lanham, MD, 9(1).

Epstein, J., Payne, K. and Kramer, E. (2002): Techniques for Mapping Suburban Sprawl, *Photogrammetric Engineering and Remote Sensing,* 63(9): PP 913-918.

Falconer, J. (1990): Hungry Season Food from the Forests, *Unasylva,* 41: PP 160.

FAO (2006): Global Forest Resources Assessment 2005, Rome, Italy.

Foody, G. and Arora, M. (1997): An Evaluation of Some Factors Affecting the Accuracy of Classification by an Artificial Neural Network, *International Journal of Remote Sensing,* 18(4): PP 799-810.

Foody, G.M. (2002): Status of Land Cover Classification Accuracy Assessment, *Remote Sensing of Environment,* 80(1): PP 185-201.

Foody, G.M. (2010): Assessing the Accuracy of Land Cover Change with Imperfect Ground Reference Data, *Remote Sensing of Environment,* 114(10): PP 2271-2285.

Foody, G.M. et al. (1995): Classification of Remotely Sensed Data by an Artificial Neural Network: Issues Related to Training Data Characteristics, *Photogrammetric Engineering and Remote Sensing,* 61(4): PP 391-401.

Franklin, S.E. and Wilson, B. (1991): Vegetation Mapping and Change Detection Using SPOT MLA and Landsat Imagery in Kluane National Park, *Canadian Journal of Remote Sensing,* 17(1): PP 2-22.

Gajbhiye, S. and Sharma, S.K. (2012): Land Use and Land Cover Change Detection of Indra River Watershed through Remote Sensing Using Multi-Temporal satellite Data, *International Journal of Geomatics and Geosciences,* 3(1): PP 89-96.

Gordon, S.I. (1980): Utilizing LANDSAT Imagery to Monitor Land Use Change: A Case Study in Ohio, *Remote Sensing of Environment,* 9(3): PP 89-196.

Groundwater Investigation Organisation (1987): Technical Feasibility of Groundwater Development on Compact Area Basis, Distt Sahranpur, Ground water survey division, Roorkee.

Gupta, K.M. and Desh, B. (1979): Man and Forest (A New Dimension in the Himalaya), Today and Tomorrow's Printer and Publisher, New Delhi.

Hansen, M.C. et al. (2000): Global Land Cover Classification at 1 km Spatial Resolution Using a Classification Tree Approach, *International Journal of Remote Sensing,* 21(6-7): PP 1331-1364.

Harding, R.A. and Scot, R.B. (1978): Forestry inventory with Landsat, Phase II Washington for productivity study, Deptt of Natural Resources, Olympia, Washington.

Hart, W.E. (1932): Forest of Madurai District, *Journal of Madras Geographical Association*, XII.

Hayes, D.J. and Sader, S.A. (2001): Comparison of Change-Detection Techniques for Monitoring Tropical Forest Clearing and Vegetation Regrowth in a Time Series, *Photogrammetric Engineering and Remote Sensing*, 67(9): PP 1067-1075.

Hilderbrandt, G. (1986): Potential and Limitation of Remote Sensing for Forestry Inventory and Mapping, Bonn, PP 165-185.

Himachal Pradesh Forest Statistics (2002): Department of Forest Farming and Environmental Conservation, H.P. Shimla.

Hoskins, M. (1990): The Contribution of Forestry to Food Security, *Unasylva*, 42: PP 160.

Howard, J.A. (1976): Remote Sensing of Tropical Forests with Special Reference to Satellite Imagery, Proceedings of Symposium Oslo, Norway.

Howarth, P.J. and Wickware, G.M. (1981): Procedures for Change Detection Using Landsat Digital Data, *International Journal of Remote Sensing*, 2(3): PP 277-291.

Jadhav, H.V. (1997): Energy and Environment, Himalayan Publishing House, New Delhi.

Jahari, M. et al. (2011): Change Detection Studies in Matang Mangrove Forest Area, Perak, *Pertanika Journal of Science and Technology*, 19(2): PP 307-327.

Jensen, J. et al. (1987): Inland Wetland Change Detection Using Aircraft MSS Data, *Photogrammetric Engineering and Remote Sensing*, 53(5): PP 521-529.

Jensen, J. et al. (1993): An Evaluation of the Coast Watch Change Detection Protocol in South Carolina, *Photogrammetric Engineering and Remote Sensing*, 59(6): PP 1039-1044.

Jensen, J. et al. (1999): Predictive Modelling of Coniferous Forest Age Using Statistical and Artificial Neural Net- work Approaches Applied to Remote Sensor Data, *International Journal of Remote Sensing*, 20(14): PP 2805-2822.

Jensen, J.R. (2005): Introductory Digital Image Processing: A Remote Sensing Perspective, In: K. C. Clarke, Ed., 3rd Edition, Prentice Hall, The United States of America.

Jensen, J.R. and Toll, D. (1982): Detecting Residential Land Use Development at the Urban Fringe, *Photogrammetric Engineering and Remote Sensing,* 48(4): PP 629-643.

Jensen, J.R., Qiu, F. and Patterson, K. (2001): A Neural Network Image Interpretation System to Extract Rural and Urban Land Use and Land Cover Information from Remote Sensor Data, *Geocarto International,* 16(1): PP 21-30.

Jensen, J.R.E. (1983): Urban/Suburban Land Use Analysis, *American Society of Photogrammetry,* Falls Church, Virginia, 2: PP 1571-1666.

Karteris, M.A. (1990): The Utility of Digital Thematic Mapper Data for Natural Resources Classification, *International Journal of Remote Sensing,* 11(9): PP 1589-1598.

Kellogg, C.E. (1961): Soil Interpretation in the Soil Survey, *Soil Conservation,* Washington, U.S. Deptt Agrik.

Kepner, W.G. et al. (2000): A Landscape Approach for Detecting and Evaluating Change in a Semi-Arid Environment, *Environmental Monitoring and Assessment,* 64(1): PP 179-195.

Khosala, P.K. and Sehgal, R.N. (1988): Trends in Tree Sciences, *Indian society of Tree Sciences,* Solan.

Khoshoo, T.N. (1988): Environmental Concerns and Strategies, Ashish Publishing House, New Delhi.

Khosla, P.K. (1992): Status of Indian Forestry Problems and Prespectives, Nirmal Vijay Printer, New Delhi.

Kochhar, P.L.(1967): Plant Ecology, Genetics and Evolution, Atma Ram and Sons, New Delhi.

Kuemmerle, T. et al. (2006): Cross-Border Comparison of Land Cover and Landscape Pattern in Eastern Europe Using a Hybrid Classification Technique, *Remote Sensing of Environment,* 103(4): PP 449-464.

Lal, J.B., Gulati, A.K. and Bist, M.S. (1991): Satellite Mapping of Alpine Pastures in the Himalaya, *International Journal of Remote Sensing,* 12(3): PP 438-443.

Latifovic, R. and Olthof, I. (2004): Accuracy Assessment Using Sub-Pixel Fractional Error Matrices of Global Land Cover Products Derived from Satellite Data, *Remote Sensing of Environment,* 90(2): PP 153-165.

Li, X. and Yeh, A. (1998): Principal Component Analysis of Stacked Multi-Temporal Images for the Monitoring of Rapid Urban Expansion in the Pearl River Delta, *International Journal of Remote Sensing,* 19(8): PP 1501-1518.

Liu, C. et al. (2007): Comparative Assessment of the Measures of Thematic Classification Accuracy, *Remote Sensing of Environment*, 107(4): PP 606-616.

Liu, H. and Zhou, Q. (2004): Accuracy Analysis of Remote Sensing Change Detection by Rule-Based Rationality Evaluation with Post-Classification Comparison, *International Journal of Remote Sensing*, 25(5): PP 1037-1050.

Liu, X and Lathrop, R.G. (2002): Urban Change Detection Based on an Artificial Neural Network, *International Journal of Remote Sensing*, 23(12): PP 2513-2518.

Liu, Y. et al. (2004): Analysis of Four Change Detection Algorithms in Bi-Temporal Space with a Case Study, *International Journal of Remote Sensing*, 25(11): PP 2121-2139.

Lu, D. et al. (2004): Change Detection Techniques, *International Journal of Remote Sensing*, 25(12): PP 2365-2401.

Lucas, R.M. et al. (1993): Characterizing Tropical Secondary Forests Using Multi-Temporal Landsat Sensor Imagery, *International Journal of Remote Sensing*, 14(16): PP 3061-3067.

Macleod, R.D. and Congalton, R.G. (1998): A Quantitative Comparison of Change-Detection Algorithms for Monitoring Eelgrass from Remotely Sensed Data, *Photogrammetric Engineering and Remote Sensing*, 64(3): PP 207-216.

Majumdar, R.C. and Pusalker, A.D. (1960): The History of Culture of the Indian People, I-IV, Bombay.

Manonmani, R. and Mary, D.G. (2010): Remote Sensing and GIS Application in Change Detection Study in Urban Zone Using Multi Temporal Satellite Data, *International Journal of Geomatics and Geosciences*, 1(1): PP 60-65.

Mas, J.F. (1999): Monitoring Land Cover Changes: A Comparison of Change Detection Techniques, *International Journal of Remote Sensing*, 20(1): PP 139-152.

Mas, J.F. (2004): Mapping Land Use/Cover in a Tropical Coastal Area Using Satellite Sensor Data, GIS and Artificial Neural Networks, *Estuarine, Coastal and Shelf Science*, 59(2): PP 219-230.

Michalak, W.Z. (1993): GIS in Land Use Change Analysis: Integration of Remotely Sensed Data into GIS, *Applied Geography*, 13(1): PP 28-44.

Michalek, J.L. et al. (1993): Multispectral Change Vector Analysis for Monitoring Coastal Marine Environments, *Photogrammetric Engineering and Remote Sensing*, 59(3): PP 635-641.

Miettinen, J. et al. (2011): Land Cover Map of Insular Southeast Asia in 250-m Spatial Resolution, *Remote Sensing Letters,* 3(1): PP 11-20.

Mouat, D.A. et al. (1993): Remote Sensing Techniques in the Analysis of Change Detection, *Geocarto International,* 8(2): PP 39-50.

Muchoney, D.M. and Haack, B.N. (1994): Change Detection for Monitoring Forest Defoliation, *Photogrammetric Engineering and Remote Sensing,* 60(10): PP 1243-1251.

Munyati, C. (2000): Wetland Change Detection on the Kafue Flats, Zambia, by Classification of a Multitemporal Remote Sensing Image Dataset, *International Journal of Remote Sensing,* 21(9): PP 1787-1806.

Nag, P., Chandra, S.K. and Sengupta, S. (2001): Environment, Population and Development, Concept Publication. Co. New Delhi.

Nagarajan, N. and Poongothai, S. (2012): Effect of Land Use/ Land Cover Change Detection of Ungauged Watershed, *World Applied Sciences Journal,* 17(6): PP 718-723.

National Remote Sensing Agency (2006): Manual of National Land use/Land cover Mapping using Multi-Temporal Satellite Imagery, Part-I, NRSA, Hyderabad.

Nelson, R. (1983): Detecting Forest Canopy Change Due to Insect Activity Using Landsat MSS, *Photogrammetric Engineering and Remote Sensing,* 49, PP 1303-1314.

Olofsson, P.P. et al. (2013): Making Better Use of Accuracy Data in Land Change Studies: Estimating Accuracy and Area and Quantifying Uncertainty Using Stratified Estimation, *Remote Sensing of Environment,* 129(15): PP 122-131.

Oort, P.A.J. (2007): Interpreting the Change Detection Error Matrix, *Remote Sensing of Environment,* 108(1): PP 1-8.

Oosting, J.H. (1956): The Study of Plant Community, W. H. Freeman and Company, London.

Oudum (1963): Ecology, Amerind Publishing Co. Pvt. Ltd., New Delhi.

Pandey, D.P. (1998): Beyond Vanishing Woods, Participatory Survival Options for Wild Life, forests and people, Himanshu Publication Udaipur-New Delhi.

Pant, D.N. and Roy, P.S. (1990): Vegetation and Land Use Analysis of Aglar Watershed Using Satellite Remote Sensing Technique, *Journal of Indian society of remote sensing,* 18(4): PP 1-14.

Peijun, D. et al. (2012): Fusion of Difference Images for Change Detection over Urban Areas, *IEEE Journal of Selected Topics in Applied Earth Observations and Remote Sensing,* 5(4): PP 1076-1086.

Pelorosso, R. et al. (2009): Land Cover and Land Use Change in the Italian Central Apennines: A Comparison of Assessment Methods, *Applied Geography*, 29(1): PP 35-48.

Peters, A.J. et al. (2002): Drought Monitoring with NDVI-Based Standardized Vegetation Index, *Photogrammetric Engineering and Remote Sensing*, 68(1): PP 71-75.

Pilon, P. et al. (1988): An Enhanced Classification Approach to Change Detection in Semi-Arid Environments, *Photogrammetric Engineering and Remote Sensing*, 54(12): PP 1709-1716.

Podeh, S.S. et al. (2009): Forest Change Detection in the North of Iran Using TM/ETM+ Imagery, *Asian Journal of Applied Sciences*, 2(6): PP 464-474.

Porwal, M.C. (1997): Remote Sensing Analysis of Environmental Resources for Planning and Development, APH Publishing Co., New Delhi.

Prakasam, C. (2010): Land use and Land cover Change Detection through Remote Sensing Approach: A Case Study of Kodaikanal Taluk, Tamil Nadu. Department of Geography, The University of Burdwan, *International Journal of Geomatics and Geosciences*, 1(2): PP 150-158.

Prakash, A. and Gupta, R. (1998): Land-Use Mapping and Change Detection in a Coal Mining Area-A Case Study in the Jharia Coalfield, India, *International Journal of Remote Sensing*, 19(3): PP 391-410.

Puri, G.S (1950): The importance of geology in the study of vegetation, Proceeding Vol. 37th I. S. C.: Part III science congress abstract, PP 36-64

Puri, G.S. (1960): Indian Forest Ecology, Vol-I and II, Oxford Book and Stationary, New Delhi.

Qiu, F. and Jensen, J. (2004): Opening the Black Box of Neural Networks for Remote Sensing Image Classification, *International Journal of Remote Sensing*, 25(9): PP 1749-1768.

Ramachandra, T. and Kumar, U. (2004): Geographic Resources Decision Support System for Land Use, Land Cover Dynamics Analysis, *Proceedings of the FOSS/GRASS Users Conference*, Bangkok.

Rao, D.P. (1991): IRS IA Application for Land Use / Land Cover Mapping in India, *Current Science*, PP 153-167.

Reis, S. (2008): Analyzing Land Use/Land Cover Changes Using Remote Sensing and GIS in Rize, North-East Turkey, *Sensors*, 8(10): PP 6188-6202.

Ridd, M.K. and Liu, J. (1998): A Comparison of Four Algorithms for Change Detection in an Urban Environment, *Remote Sensing of Environment,* 63(2): PP 95-100.

Rogan, J. and Chen, D.M. (2004): Remote Sensing Technology for Mapping and Monitoring Land Cover and Land Use Change, *Progress in Planning,* 61(4): PP 301-325.

Rogan, J. et al. (2007): Integrating GIS and Remotely Sensed Data for Mapping Forest Disturbance and Change, In: M.A. Wulder and S. E. Franklin, Eds., *Understanding Forest Disturbance and Spatial Pattern: Remote Sensing and GIS Approaches,* PP 133-171.

Rogerson, P.A. (2002): Change Detection Thresholds for Remotely Sensed Images, *Journal of Geographical Systems,* 4(1): PP 85-97.

Rosin, P.L. and Ellis, T. (1995): Image Difference Threshold Strategies and Shadow Detection, *Proceedings of the* 6th *British Machine Vision Conference,* Citeseer.

Roy, S.B. (1995): Enabling Environment for Joint Forest Management, Inter- India Publications, New Delhi.

Saxena, N.C. (1995): Forests People and Profit, New Equation for Sustainability, Natraj Publication, Dehradun.

Schoppmann, M.W. and Tyler, W.A. (1996): Chernobyl Revisited: Monitoring Change with Change Vector Analysis, *Geocarto International,* 11(1): PP 13-27.

Schulz, J.J. et al. (2010): Monitoring Land Cover Change of the Dryland Forest Landscape of Central Chile (1975- 2008), *Applied Geography,* 30(3): PP 436-447.

Seeland, K. and Franz, S. (2000): Man in the forest; Local Knowledge and sustainable management of forests and Natural Resources in Tribal Communities in India, D.K. Print World (P) Ltd. New Delhi.

Serra, P.P. et al. (2008): Land-Cover and Land-Use Change in a Mediterranean Landscape: A Spatial Analysis of Driving Forces Integrating Biophysical and Human Factors, *Applied Geography,* 28(3): PP 189-209.

Sexton, J.O. et al. (2013): Long-Term Land Cover Dynamics by Multi-Temporal Classification across the Landsat-5 Record, *Remote Sensing of Environment,* 128(21): PP 246-258.

Sharma, P.K. (1991): Forest Resources and Their Utilization in India, Mittal Publication, New delhi.

Sharma, V.K. (1998): Trees and Environment, APH Publishing Corporation, New Delhi.

Silapaswan, C. et al. (2001): Land Cover Change on the Seward Peninsula: The Use of Remote Sensing to Evaluate the Potential Influences of Climate Warming on Historical Vegetation Dynamics, *Canadian Journal of Remote Sensing,* 27(5): PP 542-554.

Singh, A. and Harrison, A. (1985): Standardized Principal Components, *International Journal of Remote Sensing,* 6(6): PP 883-896.

Singh, C. (2000): India's Forest Policy and Forest Laws, Natraj Publisher, New Delhi.

Singh, G.B. (1987): Forest Ecology of India, Rawat Publications, Jaipur.

Singh, J. and Dhillon, S.S. (1987): Agricultural Geography, Tata McGraw-Hill Publishing Company, Ltd. New Delhi.

Singh, J.S. (1983): Environmental Regeneration in Himalayan: Concept and Strategies, Gyanodaya Prakashan, Nanital, Uttarakhand.

Singh, R.L. (1971): India, A Regional geography, Silver Jubilee Publication, *National Geographical Society of India,* Varanasi.

Sivakumar, R., Ramalingam, K., Nithyan, K.R. and Rajesh, M.G. (2004): Land Resources Evaluation using Remote Sensing Technique- A Case Study in Thiruvallur, Thiruvallur District, Tamil Nadu, *Mining Engineers Journal,* 6(1): PP 27-30.

Smith, G.H. (1969): Conservation of Natural resources, (3rd ed), John Wiley and Sons, Inc., New York.

Sohl, T.L. (1999): Change Analysis in the United Arab Emirates: An Investigation of Techniques, *Photogrammetric Engineering and Remote Sensing,* 65(4): PP 475-484.

Song, C. et al. (2001): Classification and Change Detection Using Landsat TM Data: When and How to Correct Atmospheric Effects? *Remote Sensing of Environment,* 75(2): PP 230-244.

Stamp, L.D. (1968): A Glossary of Geographical Terms (2nd ed), Longmans, London.

Sumathi, M., Kumaraswamy, K., Thyagarajan, M. and Punithavathi, J. (2011): An Analysis on Land use/Land cover Using Remote Sensing Techniques-A Case Study of Pudukkottai District, Tamilnadu, India, *International Journal of Current Research,* 3(6): PP 304-307.

Taylor, J. et al. (2000): Monitoring Landscape Change in the National Parks of England and Wales Using Aerial Photo Interpretation and GIS, *International Journal of Remote Sensing,* 21(13-14): PP 2737-2752.

Thapa, R.B. and Murayama, Y. (2009): Urban Mapping, Accuracy and Image Classification: A Comparison of Multiple Approaches in Tsukuba City, Japan, *Applied Geography*, 29(1): PP 135-144.

Thayn, J.B. (2012): Assessing Vegetation Cover on the Date of Satellite-Derived Start of Spring, *Remote Sensing Letters*, 3(8): PP 721-728.

Tiwari, B.K. and Singh, S. (1995): Ecorestoration of Degraded Hill, Kaushal publication, Shillong.

Tiwari, D.N. (1986): Forestry in National Development, Jugal Kishor and Co., Dehradun.

Tiwari, P.R. and Raj, G. (1995): Ecorestoration of Degraded Hills, Kaushal Publication, Shillong.

Trewartha, G.T. (1953): A Case for Population Geography, *Annals of the Association of American Geographer*, 43.

Trewartha, G.T. (1969): A Geography of Population, World Patterns, John Wiley and Sons, New York.

Trewartha, G.T. (1976): A Geography of Population, *Population Geography, A Reader*, McGraw-Hill, New York.

Tripathi, D. and Kumar, M. (2012): Remote Sensing Based Analysis of Land Use/Land Cover Dynamics in Takula Block, Almora District (Uttarakhand), *Journal of Human Ecology*, 38(3): PP 207-212.

Tso, B. and Mather, P.P. (2009): Classification Methods for Remotely Sensed Data, 2nd Edition, CRC Press, Boca Raton.

Vimal, O.P. and Tyagi, P.D. (1986): Fuelwood from Wasteland, New Delhi.

Wald, L. (1999): Some Terms of Reference in Data Fusion, *IEEE Transactions on Geoscience and Remote Sensing*, 37(3): PP 1190-1193.

Wang, F. (1993): Knowledge-Based Vision System for Detecting Land Changes at Urban Fringes, *IEEE Transactions on Geoscience and Remote Sensing*, 31(1): PP 136-145.

Weismiller, R. et al. (1977): Change Detection in Coastal Zone Environments, *Photogrammetric Engineering and Remote Sensing*, 43(12): PP 1533-1539.

Weng, Q. (2001): A Remote Sensing GIS Evaluation of Urban Expansion and Its Impact on Surface Temperature in the Zhujiang Delta, China, *International Journal of Remote Sensing*, 22(10): PP 1999-2014.

Whyte, R.O. (1961): Evoluation of Landuse in South-Western Asia, in L.D. Stam (ed.), A History of Landuse in arid Region, UNESCO.

Williams, D. and Stauffer, M. (1978): Monitoring Gypsy Moth Defoliation by Applying Change Detection Techniques to Landsat Imagery, *Proceeding of the Symposium on Vegetation Damage Assessment, American Society of Photogrammetry,* Falls Church, PP 221-229.

Wilson, E.H. and Sader, S.A. (2002): Detection of Forest Harvest Type Using Multiple Dates of Landsat TM Imagery, *Remote Sensing of Environment,* 80(3): PP 385-396.

Xu, M. et al. (2005): Decision Tree Regression for Soft Classification of Remote Sensing Data, *Remote Sensing of Environment,* 97(3): PP 322-336.

Yang, X. et al. (2012): Impacts of Land Use and Land Cover Changes on Evapotranspiration and Runoff at Shalamulun River Watershed, China, *Hydrology Research,* 43(1-2): PP 23-37.

Zelinskey, W. (1966): A Prologue to Population Geography, Prentice Hall, Inc. Englewood Cliff, N.J.

Zeng, Y. et al. (2010): Image Fusion for Land Cover Change Detection, *International Journal of Image and Data Fusion,* 1(2): PP 193-215.

Zhang, Q. et al. (2002): Urban Built-up Land Change Detection with Road Density and Spectral Information from Multi-Temporal Landsat TM Data, *International Journal of Remote Sensing,* 23(15): PP 3057-3078.

Zhang, Y. et al. (2007): Hybrid Change Detection for Watershed Impervious Surface Using Multi-Time Remotely Sensed Data, *IEEE International of Geoscience and Re- mote Sensing Symposium,* Barcelona, PP 1939-1942.

Zheng, D. et al. (1997): Rates and Patterns of Landscape Change between 1972 and 1988 in the Changbai Mountain Area of China and North Korea, *Landscape Ecology,* 12(4): PP 241-254.

Zhou, W. et al. (2009): Object-Based Land Cover Classification of Shaded Areas in High Spatial Resolution Imagery of Urban Areas: A Comparison Study, *Remote Sensing of Environment,* 113(8): PP 1769-1777.

Zimmerman, E.W. (1951): World Resource and Industries (2nded), Harper, New York.

Zimmerman, P.P.L. et al. (2013): An Accuracy Assessment of Forest Disturbance Mapping in the Western Great Lakes, *Remote Sensing of Environment,* 128(21): PP 176-185.

APPENDIX

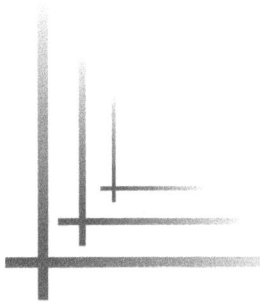

Human-Forest Interaction in Renuka Forest Division of Himachal Pradesh: A Geographical Analysis

Questionnaire

A.   **RESPONDENT'S GENERAL INFORMATION**

1. Name_____   **2.** Age_____   **3.** Sex (M/F)_____

4. Educational Qualifications_____   **5.** Occupation/Designation_____

6. Department (if working)_____   **7.** Local Address_____

B.   **PEOPLE'S PERCEPTION ABOUT ENVIRONMENT AND FOREST COVER**

1. How do you define the environment? _____

2. In your opinion, how important is it to protect the environment?
i) Very Important   ii) Important   iii) Neither Important nor Unimportant
iv) Unimportant

3. What is the role of forests in protecting environment?_____

4. What is the position of forest cover in and around your neighbourhood/ village.
i) Dense   ii) Open (Sparse)   iii) Bushes   i) No forest

5. What are the types of forests, found in your neighbourhood_____

6. Do you feel that the forest cover has changed in past few decades? Specify the change
i) Forest cover increased ii) Decreased iii) Not changed iv)Types of forests have changed v) Biomass changed

7. What was the impact of this change on environment?
i) Increase in temperature ii) Rainfall decreased iii) Landsliding
iv) Loss of wildlife v) Any other

8. What was the impact of recent development like roads, constructions of buildings, electricity etc. on the forests_____

9. What is your opinion about the forests?
i) Should be cut to earn money ii) Should not be disturbed iii) Should be raised more iv) Should be cleared for agricultural land v) Any other specify
10. How many livestock do you have?

	Type of animals	Number
1.	Cattle	
2.	Sheep	
3.	Goats	
4.	Horses	
5.	Other (specify)	

11. What are the ways of livestock grazing?
i) Stall feeding ii) Open Grazing iii) Partly both v) Misc.

12. What are the greatest threats to our green cover?
i) Forest fires ii) Livestock's grazing iii) Human beings iv) Forest smugglers
v) Construction of roads/buildings vi) Mining activity

13. Do you depend on forest land or forest resources for your livelihood?
i) Yes, very much ii) Yes iii) Not much iv) Not at all v) Other (specify)

B. PEOPLE'S PERCEPTION ABOUT LAND USE/ LAND COVER

1. What do you mean by land use?
i) Land under agricultural uses ii) Mining activities iii) Horticulture practices
iv) Any other

2. What is your opinion about land cover?
i) Land under forest and vegetation ii) Barren and rocky land
iii) Grass and grazing land iv) Any other

3. What is the current land use pattern in the Renuka forest division?
i) The area under forest cover is low
ii) Area under permanent pasture and other grazing land is quite high in the division
iii) Area under fallow land is negligible
iv) Land put to non-agricultural uses is quite low

4. What kind of forest cover changes you have experienced in your life time in the region?
i) Forests have depleted ii) Forests have degraded iii) Area under forest reduced iv) Forest area has gone up

C. PEOPLE'S PERCEPTION ABOUT FOREST SECTOR POLICIES

1. Do you know the legal rules about forest management and use?
i) Fully know ii) Partially know iii) Little idea iv) Not sure v) Do not know at all

2. Do you think these rules are clear to all?
i)Yes, they are all clear ii) Some rules are clear iii) Rules are confusing/contradicting iv) No, they are not clear at all

3. What is your opinion about T.D. rights?
i) Should continue as it is ii) Should be more liberal iii) Restricted iv)Should be banned v) Any other

4. Do you have some idea of National and State forest policy?
i) Yes ii) Some iii) No sure iv) No

5. Would you like to participate in decision making on how forest resources are managed and used?
i) Yes ii) In some (specify which ones) iii) Other (specify) iv) No

6. In your opinion what are causes of forest fire?
i) Human Negligence ii) Deliberately iii) Natural iv) Other (specify)

7. In order to control perpetual forest fire, what should be done?
i) Legal action ii) Proper legislation iii) Awareness of community iv)Any other

8. Is unscientific mining or quarrying need to be continued in forest area?
i) Yes ii) Not sure iii) Other (specify) iv) No

9. Who should be given the task of afforestation?
i) Forest Department ii) NGO's iii) Village people iv) School children iv) Any other

10. What are the past achievements of various afforestation programmes like social forestry?_____

11. How this can be improved for future success?_____

12. What is the role of forest administration in protecting as well as increasing the forest cover?_____

13. What do you suggest to better management?_____

Thank You

(Jagdish Chand)
Research Scholar

PUBLICATIONS

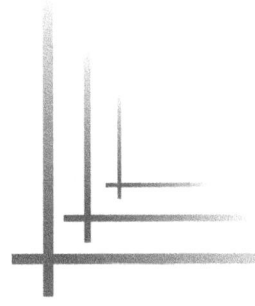

PUBLICATIONS

A) Reference Book Published:

1. Garbage Disposal System in Nahan Town, HP, India-A Case Study, International *Lap Lambert Academic Publishing* from Germany in 2013, [ISBN 9783-659-50271-2].

Research Papers Published in Refereed International Journals:

1. Changing Geography of Himachal Pradesh, Published in *International Journal of Innovative Research in Science, Engineering and Technology,* Vol.2, Issue 11, November 2013 [ISSN: 2319-8753], Impact Factor 1.6.

2. Himachal Pradesh: Trends in Urbanization, Published in *Asian Journal of Multidisciplinary Studies,* Vol.1, Issue 4, November 2013, [ISSN: 2321-8819], Indexed Journal.

3. Geographical Analysis of Kashang Hydroelectric Project in Kinnaur District, HP-A Case Study, Published in *Research Reformer, International Referred Online Research Journal,* Issue XV, December 2013, [ISSN: 2319-6904], Impact Factor 2.5.

4. Phytogeographical Analysis of Medicinal Plants in Renuka Forest Division, Published in *Research Reformer, International Referred Online Research Journal,* Issue XVI, January 2014, [ISSN: 2319-6904], Impact Factor 2.5.

5. Mechanism of Landslides of National Highway-22 in Himachal Pradesh, Published in *Grip-The Standard Research, International Referred Online Research Journal,* Issue XX, February 2014 [ISSN: 2278-8123], Impact Factor 2.5.

6. Remote Sensing and GIS based Forest Cover Change Detection Study in Renuka Forest Division, Himachal Pradesh, Published in *International Journal of Multidisciplinary Educational Research,* Vol.3, Issue 2(5), February 2014 [ISSN: 2277-7881], Impact Factor 2.735.

7. Geographical Analysis of Simbalbara Wildlife Sanctuary in Sirmour District, Published in *Research Way, International Referred Online Research Journal,* Issue XIX, February 2014, [ISSN: 2319-3557], Impact Factor 2.5.

8. Free and Open Source Software's for Geographic Information System (GIS), Published in *European Academic Research, International Multidisciplinary Research Journal,* Vol.I, Issue 12/ March 2014 [ISSN: 2286-4822], Impact Factor 3.1.

9.   Change Detection in Land Use and Land Cover Using Remote Sensing Data and GIS in Renuka Forest Division, Published in *European Academic Research, International Multidisciplinary Research Journal,* Vol.II, Issue 1/ April 2014 [ISSN: 2286-4822], Impact Factor 3.1.

Research Papers Published in Refereed National Journals:

10.   Spatio-Temporal Study of Garbage: A Case Study of Nahan Town, H.P., Published in *The Goa Geographer, The Research Journal of Geographer's Association Goa (GAG),* Vol.VII No.1 December 2010, [ISSN: 0976-786X] (Co-authored with Prof. D.D. Sharma).

11.   Impact of Hydropower Projects in Himachal Pradesh: A Case Study of Renuka Dam Project, Published in *Annals of the National Association of Geographers, India,* Vol.XXXII, No.1, June 2012, [ISSN: 0970-972X] (Co-authored with Prof. D.D. Sharma).

Participation in Seminars/Conferences:

A) Papers Presented in International Seminars/Conferences:

1.   Presented a Paper on "Socio-economic and Environmental Impact of Renuka Dam HP- A Case Study" in the *International Geographical Union (IGU) Conference* on "Geoinformatics for Biodiversity and Climate Change" Organized by Deptt. of Geography of Maharshi Dayanand University, Haryana (India) Rohtak from March 14-16, 2013.

2.   Presented a Paper on "Climate Change and Impact on Livestock of Changpa Nomadic Community in Eastern Ladakh, J&K" in the *1st International Conference* of Association of Punjab Geographers on "Disasters, Natural Resources Management and Socio-economic Development" Organized by Deptt. of Geography of Kurukshetra University, Haryana (India) from October 4-5, 2013.

3.   Presented a Paper on "Socio-economic and Environmental Impact of Hydropower-A Case Study of Bajoli Holi Hydropower" in *4th International Seminar* on "Save Rivers to Save Children of Our Children: Drawing Comprehensive Plans for River Water Management, Monitoring and Enforcement" Organized by Institute for Spatial Planning and Environment Research (ISPER), India, Panchkula from December 10-12, 2013.

B) Papers Presented in National Seminars/Conferences:

4.   Presented a Paper on "Conservation of Bio-diversity in the Himalaya-A Case Study of Kinnour District" in the *National Conference* of Association of Punjab Geographers on "Mountain Environment and Natural Resource Management" Organized by Deptt. of Geography, HPU, Shimla from October 8-9, 2011.

5. Presented a Paper on "Application of Remote Sensing in Forest Resources" in the *Regional Seminar* on "Forests for Sustainable Livelihood" Organized by Institute of Mountain Environment Bhaderwah Campus, University of Jammu on May12[th], 2012.

6. Presented a Paper on "Forest Resources of Sirmour District" in the *National Conference* on "Environmental Issues and Challenges- the Himalayan Perspective" Organized by Institute of Mountain Environment Bhaderwah Campus, University of Jammu from October 4-5, 2012.

7. Presented a Paper on "Disaster Management and Landslides- A Case Study of NH-22" in the *National Seminar* on "Natural Disasters: Vulnerability, Preparedness and Mitigation" Organized by The Environmental Forum, NSCBM, Govt. College, Hamirpur (HP) from October 18-20, 2013.

8. Presented a Paper on "Change Detection in Forest Cover Using Remote Sensing Data and GIS-A Case Study" in the *National Seminar* on "Natural Resource Management for Sustainable Development: Present Needs and Our Common Future" Organized by Youth for Sustainable Development (YSD), Shimla and Maharaja Agrasen University, Solan from March 29-30, 2014.

C) Participation in National Conferences/ Seminars:

9. Participated in *National Conference* on Sustainable Use of Water Resources in context of Climate Change: A Shared Responsibility from 11-13 March 2011, Organized by Govt. PG College, Chamba.

D) Participation in Training and Workshops

10. Participated in Special Training on *"High Resolution Image Analysis for Natural Hazard Assessment"* Organised by *Indian Institute of Remote Sensing* (ISRO) at Dehradun from January 2 to 20, 2012.

11. Participated in Training Workshop on *"Open Source Geospatial Tools"* Organized by *Centre for Space Science and Technology Education in Asia and the Pacific* (affiliated to the United Nations) and *Indian Institute of Remote Sensing* (ISRO) and *Open Source Geospatial Foundation of India* (OSGeo-India) at Dehradun from April 2 to 4, 2012.

12. Participated in *"Induction Training Programme"* Organised by *Govt. College of Teacher Education at Dharamshala* from December 24, 2012 to January 5, 2013.

13. Online Course on *"Climate Change"* from *Macquarie University* by Open Universities Australia from 24 March to 19 April 2014.

International Journal of Multidisciplinary Educational Research
ISSN : 2277-7881; Impact Factor - 2.735; IC Value:5.16
Volume 3, Issue 2(5), February 2014

REMOTE SENSING AND GIS BASED FOREST COVER CHANGE DETECTION STUDY IN RENUKA FOREST DIVISION, HIMACHAL PRADESH

Jagdish Chand
Assistant Professor
Deptt. of Geography
Govt. PG College, Nahan, India

Introduction

Land use and land cover (LULC) refers to the physical characteristics of earth surface, captured in the distribution of vegetation, water, soil and other physical features of the land, including those created solely by human activities (Louisa and Antonio, 2001). Information on the land use and land cover in the form of maps and data is very important for planning, management, and utilization of land for agriculture, forestry, urban, industrial, environmental studies and economic development (Roy and Giriraj, 2008). Assessing and monitoring the state of the earth surface is a key requirement for global change research (Jung et al., 2006; Lambin et al., 2001). With the impending threat to environment, vegetation cover mapping is now being given the highest priority. Classifying and mapping of vegetation is an important technical task for managing natural resources as vegetation provides a base for all living beings and plays an essential role in affecting global climate change, such as influencing terrestrial CO_2 (Xiao et al., 2004). Vegetation mapping also presents valuable information for understanding the natural and man-made environments through quantifying vegetation cover from local to global scales at a given time point or over a continuous period. It is critical to obtain current states of vegetation cover in order to initiate vegetation protection and restoration programmes (Egbert et al., 2002; He et al., 2005).

EUROPEAN ACADEMIC RESEARCH
Vol. I, Issue 12/ March 2014

ISSN 2286-4822
www.euacademic.org

Impact Factor: 3.1 (UIF)
DRJI Value: 5.9 (B+)

Free and Open Source Software's for Geographic Information System (GIS)

JAGDISH CHAND
Department of Geography
Govt. PG College, Nahan, HP
India

Abstract:

 Open source may be viewed by many as a revolutionary phenomenon that is capable of providing the software industry with an alternative and competitive way of doing business. Research done so far has tackled the history and business aspects of the open source phenomena, and only few have researched its technical aspects. The results of the research provide an insight on how different categories of people view open source, and demonstrate that lack of awareness about open source concepts and its competencies may be a major reason behind the poor adoption of open source solutions. The results of the comparative analysis also demonstrate that Map Server is technically equivalent to its commercial counter parts. A new open source sharing platform has emerged from the joint initiatives of the educationist, researchers and software developers. Present paper throws light on the contemporary development of this open source technology in field of geographic information system (GIS). An effort has been made to review the main components of open source technology at desktop and web based platforms, while the significance of open source technologies has been also discussed.

Key words: Open source; GIS; Map Server; Educationist; Open source technology.

Introduction

The development of free and open source software has

EUROPEAN
ACADEMIC
RESEARCH
ISSN 2286-4822
www.euacademic.org

EUROPEAN ACADEMIC RESEARCH
Vol. II, Issue 1/ April 2014

Impact Factor: 3.1 (UIF)
DRJI Value: 5.9 (B+)

Change Detection in Land Use and Land Cover Using Remote Sensing Data and GIS in Renuka Forest Division

JAGDISH CHAND
Department of Geography
Govt. PG College, Nahan
India

Abstract:

Land use and land cover change have been among the most important perceptible changes taking place around us. Although perceptible, the magnitude, variety and the spatial variability of the changes taking place has made the quantification and assessment of land use and land cover changes a challenge to geographers. Furthermore, since most of the land use and land cover changes are directly influenced by human activities. The Remote Sensing and Geographic Information System has proved to be very important in assessing and analyzing land use and land cover changes. Satellite-based Remote Sensing, by virtue of its ability to provide synoptic information of land use and land cover at a particular time and location, has revolutionized the study of land use and land cover change. The temporal information on land use and land cover helps identify the areas of change in a region. The use of Geoinformatics has enabled us to assign spatial connotations to land use land cover changes, namely, population pressure, climate, terrain, etc. which drive these changes. This has helped geographers to quantify these tools and to predict various scenarios. This article gives an overview of the current trends in land use and land cover changes of Renuka forest division in Sirmour district.

Key words: Land use; Land cover; Remote sensing; Geographic information system; Geoinformatics.

432

Annals, Volume XXXII, No. 1, June 2012 (ISSN: 0970-972X)

IMPACT OF HYDROPOWER PROJECTS IN HIMACHAL PRADESH
A Case Study of Renuka Dam Project

D.D. Sharma* and Jagdish Chand**

ABSTRACT

In India thee has been efforts to build mainly runof-the-river projects, multi-purpose hydropower plants with water storage facilities that has been important flood regulators has supported irrigation and has also provided much needed drinking water. Further, India's hydro-resources are largely available in the poorest part of the country. These resources posseses huge potentials for overall socio-economic development, poverty alleviation and regional development. Also hydro-power plays an important role in energy and development strategies of India and hence investing much projects is a necessity.

Environments and social impacts-potentially both positive and negative are invitable in case of hydropwer projects in India. A clear understanding of such impacts drawing inferences, both from analyitical as well as local knowledge is therefore required. The present paper is therefore an attempt to investigate the impact of Renuka Dam Project of Himachal Pradesh.

Key words: Dams; reservoirs; Kyoto mechanism; HPPCL; project affected people; hydro-electric power projects; Gharats.

INTRODUCTION

Water and energy are indispensable for human sustenance. The demand for water and energy on a global scale has been assessed to increase significantly in the future, in accordance with population growth and improved living standards. As such, the 21st century requires to secure water resources and to introduce renewable energy sources that are environmental friendly. The "Plan of Implementation" adopted at the Johannesburg Summit (September, 2002) recommended that renewable energy technologies

---
* Deptt. of Geography, HPU, Shimla, H.P., 171005
** Deptt. of Geography, Govt. P.G. College, Chamba, H.P., 176310

MAHARSHI DAYANAND UNIVERSITY
ROHTAK, HARYANA (INDIA)

This is to certify that

Prof./ Dr./ Mr./ Ms. Jagdish Chand

From Govt PG, College Chamba HP.

Participated and presented a paper entitled

Socio-economic and environmental impact

of Renuka Dam HP - a case study

In the International Geographical Union (IGU) Conference

on

GEOINFORMATICS FOR BIODIVERSITY AND CLIMATE CHANGE

Organised by
Department of Geography
March 14-16, 2013

Prof. Vladimir Kolossov
President, IGU

Dr. Mehtab Singh
Organising Secretary

Prof. M. I. Hassan
Convener

INTERNATIONAL GEOGRAPHICAL UNION (IGU)

1st International Conference

of

Association of Punjab Geographers

on

Disasters, Natural Resource Management and Socio-economic Development

October 4-5, 2013

Organized by

Department of Geography

Kurukshetra University, Kurukshetra, Haryana, India

Established by the State Legislature Act XII of 1956 (NACC Accredited 'A' Grade)

This is to certify that Prof. / Dr. / Mr. / Ms. _Jagdish Claud (Assistant Professor)_ from _Govt. P.G College , Nahau , Himachal Pradesh_ participated in the scientific deliberations and presented a paper entitled _Climate Change and Impact on livestock of Chougpa Nomadic community in Eastern Ladakh , Jammu and Kashmir_

He / (She) Chaired / Co-chaired the Plenary / Technical Session.

Prof. M.S. Jaglan
Organizing Secretary

Prof. S.P. Kaushik
Convener

Institute for Spatial Planning and Environment Research, India , Panchkula

(Registered under the Societies Registration Act, 1860)

Certificate of Participation

Certified that Jagdish Chand, Assistant Professor, Department of Geography , Govt. P.G. College, Nahan , H.P. presented the paper on Socio-Economic and Environmental Impacts of Bajoli Holi Hydroelectric Project in Chamba District- A Case Study in 4th International Seminar on "Save Rivers to Save Children of our Children: Drawing Comprehensive Plans for River Water Management, Monitoring and Enforcement" held from December 10-12, 2013 at Institute for Spatial Planning and Environment Research, India, Panchkula.

K. Surjit Singh
Secretary General

Manjit singh.
Prof. Manjit Singh
President

Indian Institute of Remote Sensing

Indian Space Research Organisation
Department of Space, Government of India

Enrolment No. 8341

Certificate

This is to certify that Mr./Ms./Dr. __Jagdish Chand__ from

__Govt. PG Collage, Chamba__ has been awarded this

certificate on having completed the third Special Course on **High Resolution Image Analysis for Natural**

Hazard Assessment conducted by this Institute from January 2 to 20, 2012.

Date : January 20, 2012
Place : Dehradun, India

Course Director

Dean (Academics)

Director, IIRS

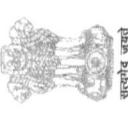

Centre for Space Science and Technology Education in Asia and the Pacific

(Affiliated to the United Nations)

Registration No. SRS-12-329

Certificate

This is to certify that Dr./Mr./Ms. __Jagdish Chand__ from

__Govt. P.G. College, Chamba__ has been awarded this certificate on having participated

the Training Workshop *Open Source Geospatial Tools* during April 2-4, 2012 conducted by this centre in collaboration

with Indian Institute of Remote Sensing (ISRO) and Open source Geospatial Foundation of India (OSGeo-India).

Minakshi Kumar
Course Coordinator

P.S. Roy
Director

Date : April 4, 2012
Dehradun

OSGeo-India

Certificate
of Achievement

jagdish chand

has successfully passed the course

Climate Change

by

April 23rd, 2014

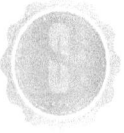

क्रम संख्या / Serial

रोल नं. / Roll No.

नामांकन संख्या / Enrolment No. 99-NN-135

........4858........

Nᵒ 0002853

हिमाचल प्रदेश विश्वविद्यालय

योग्यता प्रमाण पत्र

प्रमाणित किया जाता है कि ...जगदीश चन्द... श्री/श्रीमती/कुमारी ...सुपुत्र/सुपुत्री
श्री ...टुण्डा राम... ने इस विश्वविद्यालय से ...जून 2004...,
की ...एम.ए.(भूगोल) परीक्षा ...प्रथम... योग्यता स्थान प्राप्त करके
उत्तीर्ण की, अत: इसके प्रमाण स्वरूप इसे स्वर्ण-पदक प्रदान किया गया ।

Himachal Pradesh University

Certificate of Merit

Certified that ...Jagdish Chand..., son/daughter of
Shri ...Tunda Ram... has passed the M.A.(Geography)
examination of this University, held in ...June 2004...,
securing First class First position in the order of merit and has
been awarded a Gold Medal in token thereof.

परीक्षा नियन्त्रक / Controller of Examinations

कुल सचिव / Registrar

शिमला / Shimla 8th November 2004

क्रम संख्या / Serial

रोल नं. / Roll No.

नामांकन संख्या / Enrolment No. 99-NN-135

Nᵒ 0003076

हिमाचल प्रदेश विश्वविद्यालय

योग्यता प्रमाण पत्र

प्रमाणित किया जाता है कि ...जगदीश चन्द... श्री/श्रीमती/कुमारी ...सुपुत्र/सुपुत्री
श्री ...टुण्डा राम... ने इस विश्वविद्यालय से ...अक्तूबर 2005...,
की ...एम.फिल.(भूगोल) परीक्षा ...प्रथम... योग्यता स्थान प्राप्त करके
उत्तीर्ण की, अत: इसके प्रमाण स्वरूप इसे स्वर्ण-पदक प्रदान किया गया ।

Himachal Pradesh University

Certificate of Merit

Certified that ...Jagdish Chand..., son/daughter of
Shri ...Tunda Ram... has passed the M.Phil.(Geography)
examination of this University, held in ...October 2005...,
securing First class First position in the order of merit and has
been awarded a Gold Medal in token thereof.

परीक्षा नियन्त्रक / Controller of Examinations

कुल सचिव / Registrar

शिमला / Shimla 6th January 2006.

www.ingramcontent.com/pod-product-compliance
Lightning Source LLC
Chambersburg PA
CBHW051638170526
45167CB00001B/239